Vitamin D Deficiency - New Insights

Edited by Julia Fedotova

Published in London, United Kingdom

Vitamin D Deficiency - New Insights
http://dx.doi.org/10.5772/intechopen.100904
Edited by Julia Fedotova

Contributors
Elif Gül Aydin, Marijana Petrovic, Stanko Petrovic, Katarina Obrencevic, Patricia Loranca-Moreno, Alan Rios-Espinosa, Juan Moises Ocampo-Godínez, Teodoro Durá-Travé, Fidel Gallinas-Victoriano, Yan Zhang, Jia-Li Zhang, Yong-Jun Wang, Seema Yousuf, Sepehr Saberian, Fahim Atif, Donald Stein, Sakineh Shab-Bidar, Amin Mirrafiei, Nadia Babaee, Samira Davarzani, Mahsa Firouzi, Fatemeh Gholami, Rohollah Valizadeh, Saman Farshid, Parmida Soleimani, Pedram Ghahremani, Himangshu Sarma, Alakesh Bharali, Bhargab Deka, Bedanta Bhattacharjee, Santa Sarma, Ashique Ahmed, Suman Kumar, Susankar Kushari, Rajlakshmi Devi, Julia Fedotova

First published in London, United Kingdom, 2023 by IntechOpen
IntechOpen is the global imprint of INTECHOPEN LIMITED, registered in England and Wales, registration number: 11086078, 5 Princes Gate Court, London, SW7 2QJ, United Kingdom

British Library Cataloguing-in-Publication Data
A catalogue record for this book is available from the British Library

Additional hard and PDF copies can be obtained from orders@intechopen.com

Vitamin D Deficiency - New Insights
Edited by Julia Fedotova
p. cm.
Print ISBN 978-1-80356-515-6
Online ISBN 978-1-80356-516-3
eBook (PDF) ISBN 978-1-80356-517-0

Meet the editor

Julia O. Fedotova, MD, Ph.D. habil., Sc.D. graduated with a degree in Pharmacy from the Pharmaceutical Faculty, St. Petersburg State Chemical Pharmaceutical Academy, Russia, in 1996. She completed post-graduate training and obtained a Ph.D. in Experimental and Clinical Pharmacology and in Physiology of Humans and Animals in 1999. She completed additional training in the Department of Pharmacology, Medical School, University of Catania, Italy, in 2002, and the Institute of Physiology, Medical School, University of Pecs, Hungary in 2003-2004. She graduated with a doctorate in Neuropharmacology from the Department of Neuropharmacology, Institute for Experimental Medicine, Russian Academy of Medical Sciences, and received a Dr. habil in Biological Sciences in 2008. Dr. Fedotova is a leading researcher and principal investigator at the I.P. Pavlov Institute of Physiology, Russian Academy of Sciences and ITMO University, Russia. She has 2 grants and more than 250 publications to her credit, including 5 books and 6 book chapters. She has also edited one book. Dr. Fedotova studied the role of vitamin D in the development of affective-related disorders in women with a grant from the Russian Scientific Foundation.

Contents

Preface

Vitamin D deficiency has been noted in people worldwide and thus it is a global problem. As such, there is growing interest from scholars and health practitioners in the function of vitamin D in human health and diseases, especially in its pleiotropic outcomes. It is generally accepted that vitamin D is an essential substance for the homeostasis of calcium and phosphorus in the human body. It is a neuroactive secosteroid that may be implicated in the development of various pathological disorders. The functionally active form of vitamin D (1.25-OH-VD 2) reveals both genomic and non-genomic effects.

The book presents new insights into the role of vitamin D deficiency in the development of various chronic diseases, including COVID-19. It is organized into four sections: "Introduction", "Vitamin D Status and Childhood", "Vitamin D Status and COVID-19", and "Vitamin D Status and Non-Hormonal/Hormonal Disorders".

After the first section, which is the introduction, an important theme developed in Section 2 is vitamin D status in the neonatal and postnatal periods in children. Chapters in this section examine the consequences of vitamin D deficiency in children and discuss strategies for addressing this deficiency. Section 3 discusses the profound role of vitamin D deficiency in the era of COVID-19. Finally, Section 4 reviews the implication of vitamin D deficiency in the physiological and biochemical processes of renal, endocrine, and cardiovascular diseases.

Julia Fedotova, MD, Ph.D., Dr. habil, Sc.D.
Associated Professor,
Laboratory of Neuroendocrinology,
Pavlov Institute of Physiology RAS,
St. Petersburg, Russia

National Research South Ural State University,
Chelaybinsk, Russia

Section 1

Introduction

Chapter 1

Introductory Chapter: Vitamin D Deficiency

Julia Fedotova

1. Introduction

Vitamin D (VD) is a unique bio-regulatory molecule as it can be synthesized in the skin in addition to its dietary sources [1, 2]. The well-studied function of VD is associated with its ability to regulate metabolic processes in skeletal tissue by affecting mineralization, maintaining a balance between the formation and resorption of bone tissue, thereby contributing to the prevention of osteoporosis and the occurrence of fractures [1–3].

VD comes in two major forms, D_2 and D_3, with D_3 being the most prominent. Both forms can be sourced from food, D_2 from plants and mushrooms while D_3 can be found in fish oil [2–4]. However, the majority of the body's VD is synthesized de novo in the skin after exposure to ultraviolet B radiation such as sunlight [2, 5].

Earlier studies have suggested that a serum level of VD should range around 40–60 ng/l lowers the risks of developing different diseases, as well as an all-cause lower mortality [1, 2]. A daily intake or production of 4000–6000 international units is required to sustain these levels [2]. While it is possible to produce the required levels of VD through 2 times in a week sun exposure for 10–30 minutes, many patients are unable to produce that much because of issues related to the environment, health and socioeconomic reasons [1, 2].

That is why, it is recommended for healthy adults at risk for VD deficiency to supplement with 1500–2000 IU with an upper limit of 10.000 IU a day orally [1, 2]. A correlation between very low VD_3 levels and numerous diseases and a correlation between an impact of VD levels and normal functioning of the whole organism have well-established [1].

2. COVID-19 era and Vitamin D deficiency

VD is postulated to impact innate and adaptive immunological responses [6]. Low concentrations of VD are associated to elevated autoimmunity and increased susceptibility to virus diseases [6, 7]. The implication of VD in the protective mechanisms against respiratory tract infections by COVID-19 has been reported [7]. VD produces anti-microbial peptides secretion, especially cathelicidins and defensins, thereby resulting in the replication rate of viruses and pro-inflammatory cytokines levels. Supplementation of VD in patients with COVID-19 for treatment has been made in human studies [6, 7]. Beneficial effects of VD to protect the incidence of influenza A are associated with very low VD levels due to lack of sunlight exposure [7]. VD might help in preventing influenza by preventing a cytokine storm in the influenza state, reducing the production of IL-1 and IL-6 [7, 8]. Recently, low VD levels are linked with elevated IL-6 levels in patients with the human immunodeficiency virus disease [9].

As mentioned, VD has multiple effects on the immune system. Many studies have therefore looked into whether VD status and VD supplementation could lower the risk of contracting acute respiratory infections in general, also in some another pandemic like COVID-19 in the future.

3. Summary

Overall, VD deficiency might affect the treatment response and remission rate in different population with numerous disorders. We do not exactly know the precise mechanisms how VD deficiency might implicate in many disorders. However, we must take into account that patients with VD deficiency can be resistant to available therapeutic options, which poses a major therapeutic challenge to health experts. Thus, it is possible to suggest that VD deficiency is likely a driving factor for the development of a whole number of severe diseases.

The following recommendations can be made:

1. Each individual with VD deficiency is a unique case and needs detailed evaluation to identify the prior drug response and make a correct diagnosis.

2. Assessment of risk factors for different disorders in patients with VD deficiency is equally important to guide health professionals in tailoring an appropriate management plan for such patients.

3. There are a wide variety of options for the treatment of VD deficiency; therefore, every therapeutic paradigm needs to be utilized when helping patients with VD deficiency.

4. In light of the demonstrated importance of truly adequate VD levels to the long-term outcomes of various disorders, further randomized clinical trials involving newer drugs and therapies are needed in the future.

New technologies might offer ways to treat patients with various disorders in a association with VD deficiency more effectively.

Author details

Julia Fedotova
Laboratory of Neuroendocrinology, Pavlov Institute of Physiology RAS,
Saint Petersburg, Russia

*Address all correspondence to: julia.fedotova@mail.ru

References

[1] Holick MF, Vitamin D. Deficiency. New England Journal of Medicine. 2007;**357**:266-281

[2] Charoenngam N, Holick MF. Immunologic effects of Vitamin D on human health and disease. Nutrients. 2020;**12**:7

[3] Haddad JG, Matsuoka LY, Hollis BW, Hu YZ, Wortsman J. Human plasma transport of vitamin D after its endogenous synthesis. Journal of Clinical Investigation. 1993;**91**:2552-2555

[4] Bikle DD, Siiteri PK, Ryzen E, Haddad JG. Serum protein binding of 1,25- dihydroxyvitamin D: A reevaluation by direct measurement of free metabolite levels. Journal of Clinical Endocrinology and Metabolism. 1985;**61**:969-975

[5] Bikle DD, Gee E, Halloran B, Kowalski MA, Ryzen E, Haddad JG. Assessment of the free fraction of 25-hydroxyvitamin D in serum and its regulation by albumin and the vitamin D-binding protein. Journal of Clinical Endocrinology and Metabolism. 1986;**63**:954-959

[6] Ohaegbulam KC, Swalih M, Patel P, Smith MA, Perrin R. Vitamin D supplementation in COVID-19 patients: A clinical case series. American Journal of Therapy. 2020;**27**:e485-e490

[7] Rastogi A, Bhansali A, Khare N, et al. Short term, high-dose vitamin D supplementation for COVID-19 disease: A randomised, placebo-controlled, study (SHADE study). Postgraduate Medical Journal. 2022;**98**:87-90

[8] Annweiler G, Corvaisier M, Gautier J, et al. Vitamin D supplementation associated to better survival in hospitalized frail elderly COVID-19 patients: The GERIA-COVID quasi-experimental study. Nutrients. 2020;**12**:3377

[9] Annweiler C, Hanotte B, de Grandin C, Sabatier JM, Lafaie L, Célarier T. Vitamin D and survival in COVID-19 patients: A quasi-experimental study. Journal of Steroid Biochemistry and Molecular Biology. 2020;**204**:1-6

Section 2

Vitamin D Status and Childhood

Chapter 2

Vitamin D Deficiency in Pediatric Dentistry

Elif Gül Aydin

Abstract

Vitamin D (vitD) deficiency has essential effects on general health. It is known that oral and dental health is an integral part of public health, and there is a close relationship between them. From the development and eruption stages of the teeth to the formation of caries, vitD deficiency has accepted significant effects on oral health. It is essential to understand the role of vitD deficiency in early childhood caries (ECC), which is considered one of the most critical problems, especially in pediatric patients. Low vitD levels during pregnancy have even been reported to increase ECC risk in infancy. For this reason, care should be taken to ensure that the mother's 25(OH)d level and later the child is in optimal conditions, starting from the pregnancy period, to improve the oral health status of children.

Keywords: pediatric dentistry, vitD deficiency, early childhood caries, primary teeth

1. Introduction

Today, within the framework of increasing holistic health understanding, it is seen that the interest in the relationship between oral and systemic health has improved. The relationship between dental caries and nutrition is known from when caries are defined. 25-hydroxyvitamin D (25(OH)D) <30 nmol/L is accepted as a vitamin D deficiency, and it is stated that it may be associated with poor oral health. The incidence of periodontal disease, dental caries, and enamel hypoplasia is increased with vitamin D deficiency. In addition, 25-hydroxyvitamin D or its metabolites are involved in the immunological response together with the antimicrobial peptides cathelicidin and defensins that play an immunological role.

2. Vitamin D deficiency in pediatric dentistry

Tooth caries seen in children remain its place as a significant disease that affects more than 600 million children worldwide. Early childhood caries (ECC) is defined as the presence of one or more decayed (non-cavitated or cavitated lesions), missing or filled (due to caries) surfaces in any primary tooth of a child under 6 years of age [1]. Permanent teeth under the primary teeth play a protective place to come to their appropriate places in the mouth. Due to caries in these teeth, chronic pain, infections, and nutritional problems occur in children. These problems have a negative impact on the quality of life of children and families and increase the burden of community health programs. For this reason, they should be considered an integral part of general health.

Early childhood caries, like other types of caries, is a dynamic disease that starts with biofilm, can persist in the presence of sugar, is considered multifactorial, and occurs with the deterioration of the balance between demineralization and remineralization of dental hard tissues. Tooth decay occurs due to the biological, behavioral, and psychosocial factors associated with the individual's environment. Common risk factors for ECC include improper infant feeding, exposure to sugar, poor oral hygiene, poverty, and late access to preventive dental care. ECC has common risk factors for other noncommunicable diseases associated with excessive sugar consumption, such as cardiovascular disease, diabetes, and obesity. Excessive sugar consumption causes prolonged acid production from the bacteria attached to the tooth and changes the composition of the oral microbiota and the pH of the biofilm. When this condition persists, the tooth structures demineralize [1]. Severe early childhood caries (S-ECC) is a more aggressive manifestation that frequently necessitates dental treatment under general anesthesia. Those with S-ECC often suffer from pain, sleep disturbance, behavior changes, and altered eating habits. In recent cross-sectional studies, children with S-ECC have been found to have lower vitamin D status, iron status, or anemia [2–6].

The relationship between nutrition, general health, and dental caries studies dates to the 1920s. The study by Mellanby and Pattison, published in 1928, provides the first evidence that vitamin D deficiency is associated with dental caries in children [7, 8]. A growing body of studies and evidence shows that low serum concentrations of 25-hydroxyvitamin D (25(OH)D) are associated with increased caries experience [2, 8–11]. It has even been reported that low 25(OH)D levels in the mother during pregnancy increase the risk of ECC in children [9, 10, 12].

Minerals, such as magnesium, calcium, and phosphorus, the essential structural components of the tooth, should be taken at sufficient levels with the diet. These minerals play a role by interacting with vitamins in strengthening the tooth structure. A deficiency of magnesium, calcium, and phosphorus in diet and nutrition content usually results in loose teeth and premature tooth loss. If magnesium deficiency occurs during the formation stages of teeth, delay in eruption times, enamel or dentin hypoplasia appear. Also, the alveolar bone becomes brittle, and the gingiva becomes hypertrophic. It is stated that magnesium strengthens the antimicrobial environment, reduces oral inflammation, and increases the flexibility of tooth enamel by increasing calcium absorption in the teeth [13, 14], especially vitD is related to calcium, magnesium, and zinc. Without magnesium, the immune system cannot activate vitamin D, and sufficient calcium absorption does not occur in the teeth. vitD is a hormone essential for the intestinal absorption of Calcium, Magnesium and Phosphate. Vitamin D helps regulate calcium and phosphate balance to maintain healthy bone function. Magnesium helps activate vitamin D, which helps regulate calcium and phosphate homeostasis, influencing bone growth and maintenance. There is a synergistic relationship between vitD and magnesium [13].

Several different mechanisms have been proposed to express the role of vitD among the factors reducing caries.

One of these mechanisms is to play a role in the formation, calcification, mineralization, and protection of teeth by affecting serum calcium, phosphate levels, and parathyroid hormone. The balance between calcium and phosphate levels is important for the formation, calcification, mineralization, and preservation of teeth, bone, hard tissue, maxilla, and mandible. Enamel and dentin defects-hypoplasia have been linked with hypocalcemia and hypophosphatemia [11, 15].

Dental caries and VDD (vitamin D deficiency) affect children worldwide. Changes in both enamel and dentin are observed in children with a VDD. Therefore, vitD has a significant role in forming oral hard tissue, comprising tooth enamel and dentin, and affects primary and permanent teeth development [8, 11, 16].

Vitamin D deficiency may cause defects in enamel and dentin and increase the risk of dental caries. In a systematic review of controlled clinical trials, 2827 children were included. As a result of this systematic review, the significant relationship found between vitamin D levels, and dental caries showed that vitamin D is a promising anticaries agent [4, 8, 17].

Vitamin D has a significant role in odontogenesis [11, 18]. The mechanism by which vitD excites the mineralization of tooth enamel involves binding to vitamin D receptors (VDR) expressed in both tooth and bone cells. VDR directs the transcription of several target genes, mostly defined by ameloblasts and odontoblasts [11, 15, 19]. It coordinates physiological functions by controlling calcium and phosphate metabolism, promotes growth, and induces necessary remodeling of the bones and teeth [4]. VDR stimulates the formation of structural gene products in dentin, together with calcium-binding proteins and diverse extracellular matrix proteins. The gene encoding VDR is positioned on chromosome 12q13.11 and comprises several polymorphisms [19]. The VDR gene adjusts the biological role of primary vitD metabolites; thus, having a vital role in the configuration of teeth, particularly in mineralizing dentin and enamel. Consequently, developmental deficiencies, for example, enamel hypoplasia, can result from VDD. Ultimately, vitD and VDR at the molecular level influence the tooth germ formation, supply the regulation of enamel and dentin structure and maturation, and organize the phases of dental crown growth [11, 15]. In addition to dental problems due to vitamin D deficiency, genetic polymorphism of VDR gene polymorphism has been associated with dental problems, such as external apical root resorption, periodontal diseases, dentinogenesis imperfecta, chronic periodontitis, and dental caries [20].

Tooth development and eruption are complex mechanisms involving the resorption of alveolar bone and the eruption pathway. A disorder in these processes causes persistent primary teeth and delays in permanent tooth eruption. It is stated that decreased vitamin D levels increase the rate of constant primary teeth and cause delays in eruption in permanent teeth. It was known that maternal vitamin D deficiency affects the formation and mineralization of primary teeth. It is considered among the dental effects of vitamin D deficiency, which cause tooth eruption delays in children [20].

Moreover, vitD adjusts and adapts both the innate and adaptive immune systems. The immunological role of vitD is stimulating the arrangement of some antimicrobial peptides, for example, defensins and cathelicidin (LL-37), which defend against many pathogens, counting oral bacteria [4, 11, 12, 21]. Cathelicidin (LL-37 or hCAP-18) is controlled by vitD, which has antiendotoxin and antimicrobial properties [22]. Vitamin D induces cathelicidin (LL-37), which is found in the immune system. LL-37 exhibits both antimicrobial and antiendotoxin activity. It is stated that there is an inverse relationship between dental caries and LL-37, and the concentration of LL-37 is low in children with high caries activity. Epithelial antimicrobial peptides (LL-37 is one of them) are the protectors of the oral cavity. It is stated that these antimicrobial peptides have essential roles in reducing gingivitis in oral health. Therefore, vitamin

D may be helpful in the treatment of periodontitis due to its direct effects on bone metabolism and possible anti-inflammatory effects on periodontopathogens [13].

Early tooth loss, bone resorption, and bleeding tendency increase in mineral deficiencies due to absorption disorders [13]. The chewing efficiency of individuals provides the highest nutrient intake efficiency. Prematurely lost teeth have adverse effects on diet selection and nutritional efficiency. Deterioration of nutritional efficiency also causes deficiencies in vitamin intake. The resulting vitamin deficiency also negatively affects the formation of dental hard tissue. Vitamin deficiencies in the developmental stages of teeth cause disorders in the formation of hard tissue of the tooth. Therefore, it has been stated that there is a strong positive correlation between vitamin deficiencies. Malnutrition and dental hard tissue hypoplasia increase the risk of dental caries in children during the primary dentition period [2, 23]. In addition, mineral deficiencies cause bleeding in the gums, delayed tooth eruption, periodontal disease, and destruction patterns in the alveolar bone [13, 23].

It is stated that dental health in children has started to decrease in industrialized countries due to identifying risk factors and developing preventive strategies. However, it was determined that while the prevalence of caries decreased, enamel defects began to increase. It is considered a global burden, with the majority of enamel defects rising to 38% in Western European countries [24]. Developmental enamel defects are present when the affected tooth erupts. When the tooth newly erupts, it may appear hypo mineralized, porous, and more yellow-brown opaque in color. These impacted teeth are susceptible to fracture easily, and the permanent first molars are affected quite frequently. Although these defects predispose to atypical and extensive caries, they may result in the loss of the relevant permanent teeth in the early period of life [25]. Even if the developmental enamel defects are mild, they cause pain in children due to dentin sensitivity, poor esthetic color, and caries susceptibility. Considering the effects of enamel defects on quality of life and health care utilization, a significant public health problem arises [24–26]. Developmental enamel defects are associated with the calcification processes of the teeth. Since calcifications of primary and permanent teeth occur in the early postnatal period following the intrauterine period, it is necessary to examine these periods while investigating the etiological factors. Studies indicate maternal disease, drug use during pregnancy, premature birth, birth complications, and early childhood diseases as the etiological factors of enamel defects [27]. However, these pieces of evidence are unfortunately insufficient and therefore preventable. Vitamin D plays a crucial role in enamel formation, and vitamin D deficiency is now also considered a common health challenge in westernized societies [4].

A randomized controlled study conducted with high-dose vitamin D supplementary to mothers during pregnancy and early postpartum showed an inverse relationship between high-dose vitamin D intake and the formation of enamel defects. In other words, high-dose vitamin D supplementation taken during pregnancy and early postpartum showed a 50% reducing effect on enamel defects in 6-year-old children [26]. It is on the agenda to recommend vitamin D supplementation as a primary preventive measure for enamel defects, potentially impacting dental health. Considering the crucial role of vitamin D in enamel mineralization [19, 28], the biological relationship between vitamin D supplementation and these enamel defects is considered reasonable. Vitamin D has an effect on the function of ameloblasts and the mineralization of enamel in the early stages of tooth formation. It is stated that high-dose vitamin D supplementation has a protective effect on the structural strength and development of enamel [26].

Vitamin D is also very effective, along with minerals, in protecting oral health. Vitamin D helps maintain the calcium-phosphate balance and contributes to the shaping of the bone. The beneficial effects of vitamin D on oral health are not only limited to the direct impact on tooth mineralization. Still, they are also exerted through anti-inflammatory functions and the ability to stimulate the production of antimicrobial peptides. It also has essential functions by showing anti-inflammatory effects. It is reported that with sufficient Vit-D levels, the onset and progression of caries in the tooth structure can be stopped, caries can be formed, and enamel loss can be prevented [4, 11, 13].

Early childhood caries affects not only dental health but also has severe effects on general health. Malnutrition, iron deficiency anemia, and VDD are seen in children with ECC because their nutritional status is affected [11, 17]. When the relationship between vitamin D levels and caries was evaluated, it was determined that the incidence of dental caries was higher in the children of mothers with low serum vitamin D during pregnancy, and there was a strong correlation between vitamin D levels and DMFT scores in the early childhood period of children up to 6 years of age [3]. It is stated that the incidence of caries in the permanent first molar teeth is lower in children with serum vitamin D levels greater than 50 nmol in the 10–11 years of early adolescence period [3]. Similarly, in children aged 6–17 years, it was determined that every 10 mg/ml increase in serum vitamin D resulted in a 0.66 decrease in DMFT scores. It is also expected that malnutrition and related vitamin deficiencies increase the incidence of enamel hypoplasia [2, 5, 9, 11]. According to the results of an observational study, the rate of dental caries was found to be more than three times higher in 6-year-old children of mothers with 25 OHD deficiency in the third trimester of pregnancy compared to the children of mothers with adequate 25 OHD levels in the third trimester of pregnancy [3].

Vitamin D use may have a role in the protection of caries early in life. According to meta-analysis studies, it is thought to be a promising caries prevention agent, given that Vit-D supplementation relates to a 47% reduction in caries in children [3, 17]. Considering intrauterine life, an essential and critical stage for the development of teeth, and vitamin D deficiency during pregnancy plays a crucial role in the susceptibility to enamel hypoplasia and caries [2, 11, 23].

Improving vitD levels in children from an early stage appears to be an essential task. This requires awareness of pregnancy. Pregnant women should have their vitD levels tested routinely during the first trimester of pregnancy, and the risk of VDD, VDD, and vitD ingestion should be evaluated. Prenatal vitD levels appear to influence the development of primary dentition and ECC [3, 9, 11, 23]. In addition, a study found that pregnant women's poor oral hygiene and low vitamin D levels were positively correlated with preterm birth and low birth weight [29].

Vitamin D is an essential hormone for absorbing calcium, magnesium, and phosphorus from the intestine, which is necessary for properly mineralizing bones and teeth. Although there are no applications in children, it has been shown that coating the implant surfaces with vitD during dental implant application increases osteointegration. In addition, intraperitoneal application of vitD creates positive effects (such as facilitating acceleration) on tooth movements during orthodontic treatment [13].

3. Conclusions

When evaluated from a holistic perspective, it should be kept in mind that oral health is also a part of general health. When seeking solutions for dental problems in children, the vitamin values of individuals should be considered. Increasing social awareness in the fight against ECC and evaluating vitamin D deficiency is essential.

Conflict of interest

The authors declare no conflict of interest.

Author details

Elif Gül Aydin
Faculty of Dentistry, Sakarya University, Sakarya, Turkiye

*Address all correspondence to: eligul@sakarya.edu.tr

References

[1] Pitts N, Baez R, Diaz- Guallory C, Donly KJ, Feldens CA, McGrath C, et al. Early childhood caries: IAPD Bangkok declaration. International Journal of Paediatric Dentistry. 2019;**29**:384-386

[2] Silva CC, Mendes R, Manso MDC, Gavinha S, Melo P. Prenatal or childhood serum levels of vitamin D and dental caries in paediatric patients: A systematic review. Oral Health & Preventive Dentistry. 2020;**18**:653-668

[3] Beckett DM, Broadbent JM, Loch C, Mahoney EK, Drummond BK, Wheeler B. J dental consequences of vitamin D deficiency during pregnancy and early infancy - An observational study. International Journal of Environmental Research and Public Health. 2022;**19**(4):1932

[4] Uwitonze AM, Murererehe J, Ineza MC, Harelimana EI, Nsabimana U, Uwambaye P, et al. Effects of vitamin D status on oral health. The Journal of Steroid Biochemistry and Molecular Biology. 2018;**175**:190-194

[5] Deane S, Schroth RJ, Sharma A, Rodd C. Combined deficiencies of 25-hydroxyvitamin D and anemia in preschool children with severe early childhood caries: A case-control study. Paediatrics & Child Health. 2018;**23**(3):40-45

[6] Schroth RJ, Levi JA, Sellers EA, Friel J, Kliewer E, Moffatt ME. Vitamin D status of children with severe early childhood caries: A case-control study. BMC Pediatrics. 2013;**13**(1):1-8

[7] Mellanby M, Pattison CL. The action of vitamin D in preventing the spread and promoting the arrest of caries in children. British Medical Journal. 1928;**2**:1079-1082

[8] Carvalho Silva C, Gavinha S, Manso MC, Rodrigues R, Martins S, Guimarães JT, et al. Serum levels of vitamin d and dental caries in 7-year-old children in Porto metropolitan area. Nutrients. 2021;**13**(1):166

[9] Singleton R, Day G, Thomas T, Schroth R, Klejka J, Lenaker D, et al. Association of maternal vitamin D deficiency with early childhood caries. Journal of Dental Research. 2019;**98**(5):549-555

[10] Williams TL, Boyle J, Mittermuller BA, Carrico C, Schroth RJ. Association between vitamin D and dental caries in a sample of Canadian and American preschool-aged children. Nutrients. 2021;**13**(12):4465

[11] Almoudi MM, Hussein AS, Abu Hassan MI, Schroth RJ. Dental caries and vitamin D status in children in Asia. Pediatrics International. 2019;**61**(4):327-338

[12] Seminario AL, Velan E. Vitamin D and dental caries in primary dentition. Journal of Dentistry for Children. 2016;**83**(3):114-119

[13] Uwitonze AM, Rahman S, Ojeh N, Grant WB, Kaur H, Haq A, et al. Oral manifestations of magnesium and vitamin D inadequacy. The Journal of Steroid Biochemistry and Molecular Biology. 2020;**200**:105636

[14] Najeeb S, Zafar MS, Khurshid Z, Zohaid S, Almas K. The role of nutrition in periodontal health: An update. The role of nutrition in periodontal health: An update. Nutrients. 2016;**8**(9):530

[15] Fatturi AL, Menoncin BL, Reyes MT, Meger M, Scariot R, Brancher JA, et al. The relationship between molar incisor hypomineralization, dental caries, socioeconomic factors, and polymorphisms in the vitamin D receptor gene: A population-based study. Clinical Oral Investigations. 2020;**24**(11):3971-3980

[16] Krawiec M, Marzena D. The role of vitamin D in the human body with a special emphasis on dental issues: Literature review. Dental and Medical Problems. 2018;**55**(4):419-424

[17] Hujoel PP. Vitamin D and dental caries in controlled clinical trials: Systematic review and meta-analysis. Nutrition Reviews. 2013;**71**(2):88-97

[18] Zhang X, Beck P, Rahemtulla F, Thomas HF. Regulation of enamel and dentin mineralization by vitamin D receptor. In: Comparative Dental Morphology. Vol. 13. Basel: Karger Publishers; 2009. pp. 102-109

[19] Berdal A, Papagerakis P, Hotton D, Bailleul-Forestier I, Davideau JL. Ameloblasts and odontoblasts, target-cells for 1, 25-dihydroxyvitamin D3: A review. International Journal of Developmental Biology. 2003;**39**(1):257-262

[20] Xavier TA, Madalena IR, da Silva RAB, da Silva LAB, Silva MJ, De Rossi A, et al. Vitamin D deficiency is a risk factor for delayed tooth eruption associated with persistent primary tooth. Acta Odontologica Scandinavica. 2021;*79*(8):600-605

[21] Gyll J, Ridell K, Öhlund I, Karlsland Åkeson P, Johansson I, Lif Holgerson P. Vitamin D status and dental caries in healthy Swedish children. Nutrition Journal. 2018;**17**(1):1-10

[22] Molloy EJ, Murphy NS. Vitamin D, Covid-19 and children. Irish Medical Journal. 2020;**113**(4):64

[23] Cagetti MG, Wolf TG, Tennert C, Camoni N, Lingström P, Campus G. The role of vitamins in oral health. A systematic review and meta-analysis. International Journal of Environmental Research and Public Health. 2020;**17**(3):938

[24] Kühnisch J, Heitmüller D, Thiering E, Brockow I, Hoffmann U, Neumann C, et al. Proportion and extent of manifestation of molar-incisor-hypomineralizations according to different phenotypes. Journal of Public Health Dentistry. 2014;**74**(1):42-49

[25] Kühnisch J, Kabary L, Malyk Y, Rothmaier K, Metz I, Hickel R, et al. Relationship between caries experience and demarcated hypomineralised lesions (including MIH) in the permanent dentition of 15-year-olds. Clinical Oral Investigations. 2018;**22**(5):2013-2019

[26] Nørrisgaard PE, Haubek D, Kühnisch J, Chawes BL, Stokholm J, Bønnelykke K, et al. Association of high-dose vitamin D supplementation during pregnancy with the risk of enamel defects in offspring: A 6-year follow-up of a randomized clinical trial. JAMA Pediatrics. 2019;**173**(10):924-930

[27] Silva MJ, Scurrah KJ, Craig JM, Manton DJ, Kilpatrick N. Etiology of molar incisor hypomineralization – A systematic review. Community Dentistry and Oral Epidemiology. 2016;**44**(4):342-353

[28] Hewison M, Adams JS. Vitamin D insufficiency and skeletal development in utero. Journal of Bone and Mineral Research. 2010;**25**(1):11

[29] Ferrillo M, Migliari M, Roccuzzo A, Molinero-Mourelle P, Falcicchio G, Umano GR, et al. Periodontal disease and vitamin D deficiency in pregnant women: Which correlation with preterm and low-weight birth? Journal of Clinical Medicine. 2021;**10**(19):4578

Chapter 3

Vitamin D Deficiency and Critical Care in the Neonatal Period

Pedram Ghahremani

Abstract

Neonates in critical care constitute a vulnerable group, and vitamin D status in this group is the subject of extensive research. Studies suggest that critically ill neonates and children have lower mean vitamin D levels than healthy ones, and there is evidence linking vitamin D deficiency to an increased risk of mortality, illness severity, and complications in these patients. Vitamin D deficiency in neonates and children with congenital heart disease (CHD) undergoing corrective surgical treatment has attracted particular attention. Overall, studies show high prevalence rates of vitamin D deficiency in this group. Moreover, several studies report significant associations between low vitamin D levels and unfavorable findings, such as increased requirements for vasoactive support and mechanical ventilation and prolonged ICU stays. Available data suggest vitamin D deficiency as a risk factor in neonatal and pediatric critical illness, specifically in CHD patients undergoing surgical treatment. Clinical trials have been proposed to examine the beneficial effect of preoperational vitamin D supplementation on the outcome in this group. However, for now, vitamin D supplementation should be considered in critically ill neonates, particularly those undergoing surgery for CHD, aiming to maintain vitamin D at safe levels over the threshold of vitamin D deficiency.

Keywords: vitamin D deficiency, neonatal period, critical care, congenital heart disease

1. Introduction

The traditional roles of vitamin D in skeletal dynamics, renal calcium reabsorption, and intestinal calcium absorption have long been known, as are the effects of vitamin D deficiency on skeletal abnormalities such as rickets [1], but in recent years, our understanding of vitamin D functions has rapidly evolved, and the list of roles attributed to this compound in the body is greatly expanded. Basic science studies have demonstrated the presence of vitamin D receptors on a wide range of cell types in the body, from myocytes to white blood cells, implying a physiological role for vitamin D in these organs. Similarly, laboratory animal studies have shown that a large number of diseases can occur in genetically modified, vitamin D receptor knockout mice and also in normal animals with nutritionally-induced vitamin D deficiency. In humans, vitamin D deficiency is now linked to various disorders, from unfavorable pregnancy outcomes to an increase in all-cause mortality, cardiovascular and cerebrovascular events, infectious and autoimmune diseases, and numerous cancers [2]. As a result, vitamin D is now studied as a crucial substance with potential effects on the initiation, progression, and outcome of a wide range of pathological conditions.

The possible links between vitamin D status and outcomes in critically sick patients are fascinating subjects of current research. It has been hypothesized that vitamin D deficiency may have a negative impact on the outcome in critically ill patients due to a variety of roles that this vitamin plays in vital organs and life processes [3]. Several studies have found a link between vitamin D deficiency and the risk of death in people receiving critical care, and clinical trials are conducted to look at the potential positive effect of supplemental vitamin D in this population of patients [4]. Neonates and infants in critical care constitute a particularly vulnerable group, and a considerable amount of research is focused on vitamin D status in this group. Studies suggest that neonates and children who are critically ill have lower mean vitamin D levels than healthy ones, and there is evidence linking vitamin D deficiency to an increased risk of mortality, illness severity, and complications in these patients [5, 6]. In this chapter, we look into the ongoing research on this exciting topic.

2. Defining vitamin D deficiency

Experts generally agree that 25-hydroxyvitamin D [25(OH)D] concentration should be used to evaluate vitamin D status because it captures the contribution from both diet and dermal synthesis [7], but there has been much debate regarding the suggested thresholds (cut-offs) to define vitamin D deficiency. The Endocrine Society Task Force on Vitamin D in the US has recommended that people with serum 25(OH) D levels of less than 50 nmol/L be classified as vitamin D-deficient [8], while an international expert panel assembled by the European Society for Pediatric Endocrinology has suggested serum 25-hydroxyvitamin D levels of 30–50 nmol/L as vitamin D insufficiency and levels <30 nmol/L as deficiency [9].

In addition to clinical care guidelines, the question of vitamin D deficiency has been addressed from the population health perspective. Some expert bodies tasked with developing dietary recommendations for vitamin D propose 50 nmol/L as the concentration of serum 25(OH)D that would satisfy the physiological vitamin D requirement of nearly all "normal healthy persons" [7]. The Institute of Medicine (IOM) in the US chose calcium absorption, bone mineral density (BMD), and two well-studied clinical conditions- rickets in children and osteomalacia in adults- as indicators for developing their vitamin D recommendations, known as Dietary Reference Intakes (DRI) [7]. The DRI committee developed the Recommended Dietary Allowance (RDA) based on the estimate that serum 25(OH)D concentration of 50 nmol/L (15 μg/day for those aged 1 to 70 and 20 μg/day for those over 70) would satisfy the needs of nearly all (i.e., 97.5%) of "normal healthy persons" [7]. Likewise, the Scientific Advisory Committee on Nutrition (SACN) in the UK chose musculoskeletal status (rickets, osteomalacia, falls, muscle strength, and function) for developing their vitamin D recommendations, known as Dietary Reference Values (DRV). They believe that the evidence overall suggests the risk of poor musculoskeletal health increases at serum 25(OH)D concentrations below 20–30 nmol/L [10]. Based on this reasoning, SACN chose a serum 25(OH)D target of 25 nmol/L as the "population protective level" in order to safeguard the musculoskeletal health of people in the UK throughout the year. They suggest a Reference Nutrient Intake (RNI) of 10 μg/day for those aged 4 years and older [10].

Overall, according to guidelines, serum 25(OH)D concentrations above 50 nmol/L indicate vitamin D sufficiency for the majority of people, whereas concentrations between 50 and 30 nmol/L indicate a risk of vitamin D inadequacy or deficiency for some people [7]. Even though there is not yet universal agreement on what constitutes vitamin D deficiency, it is generally accepted that we do not want people in our populations to have 25(OH)D concentrations below 25/30 nmol/L. Preventing such vitamin D deficiency is a public health priority.

3. Vitamin D deficiency in neonates and infants

Vitamin D deficiency is one of the most significant dietary deficits in children worldwide. The International Osteoporosis Foundation (IOF) estimates that vitamin D deficiency affected 84% of pregnant women and 96% of babies in 2009 [11]. A 2016 systematic review considered 25(OH)D concentrations in neonates and pregnant women worldwide from 1959 to 2014. Results showed that the mean maternal 25(OH)D concentrations ranged from 13 to 130 nmol /L, while the mean neonatal 25(OH)D concentrations ranged from 5 to 77 nmol/L. Around 54% of pregnant women and 75% of newborns had vitamin D deficiency, defined as a serum 25(OH)D concentration below 50 nmol /L, and 18% of pregnant women and 29% of newborns had severe vitamin D deficiency, defined as a serum 25(OH)D concentration below 25 nmol/L. There was a wide variation in vitamin D levels between WHO regions. In the Eastern Mediterranean region, the average maternal 25(OH)D concentration was <25 nmol /L, while it ranged from 75 to 100 nmol/L in the African region. In the Americas, the average newborn 25(OH)D concentration was >75 nmol/L, but it was <25 nmol/L in the Eastern Mediterranean. Similarly, 79% of pregnant women in the Eastern Mediterranean region had severe vitamin D deficiency. The prevalence of severe vitamin D insufficiency among pregnant women in the Americas, Western Pacific, and Europe were much lower (9%, 13%, and 23%, respectively). The prevalence of vitamin D deficiency in neonates from the South-East Asian region was very high (96%), while in the Americas, 30% of neonates were vitamin D deficient [12].

Leading causes of vitamin D deficiency in neonates include lower nutritional intake, exclusive breastfeeding when the mother has vitamin D deficiency, and decreased sun exposure due to seasonal changes. Maternal vitamin D deficiency is demonstrated to be a significant risk factor for newborns with vitamin D deficiency [11]. Neonatal 25(OH)D concentrations at delivery are generally lower than normal in babies born to mothers with deficient and insufficient vitamin D levels but not in babies of mothers with sufficient vitamin D status [13]. Infants in a disadvantaged socioeconomic position exhibit a higher rate of vitamin D deficiency. This could be explained by the likelihood of decreased calcium and vitamin D intake in these children. Likewise, preterm newborns are more susceptible to vitamin D deficiency due to diminished placental transfer, insufficient sun exposure, and lower vitamin D storage as a result of low-fat mass. Babies with nephrotic syndrome, cystic fibrosis, or malabsorption syndrome are also at higher risk of vitamin deficiency [14]. Drugs such as phenobarbital, carbamazepine, oxcarbazepine, and phenytoin can interfere with vitamin D metabolism [8]. The absorption, metabolism, or activation of vitamin D are also impacted by other medications, including corticosteroids and azole antifungals.

Secondary hyperparathyroidism, secondary hyperphosphatemia, and hypocalcemia are the main biochemical changes seen in babies with vitamin D deficiency [15].

Along with vitamin D deficiency, other factors that may contribute to hyperphosphatemia in this group include a decreased glomerular filtration rate, low intact parathyroid hormone (iPTH) levels, and renal tubular nonresponse to PTH, particularly in the first days after birth. It is possible that co-occurring metabolic bone disease—especially in preterm, small-for-gestational-age (SGA) neonates—also contributes to these biochemical alterations.

Clinically, vitamin D deficiency in infants presents as rickets in extreme situations, which can cause bowing of the knees, wrist widening, frontal bossing, spontaneous fractures, and skeletal abnormalities. In older children, it can cause short stature, muscle weakness, and pain. In addition to skeletal signs, vitamin D insufficiency can cause developmental delay, growth failure, and recurrent respiratory infections [14]. Early vitamin D insufficiency can go undetected without clinical symptoms and later proceed to florid rickets in older children if undiagnosed. In fact, in the neonatal age group, the symptoms of hypocalcemia, there may the only clinical manifestations of vitamin D deficiency.

To avoid the risk of later skeletal abnormalities and reduce pulmonary morbidity, vitamin D insufficiency must be promptly identified and treated in neonates and infants. A dosage of 400–1000 international units (IU) of vitamin D per day for 8 to 12 weeks is the recommended course of treatment for newborns with vitamin D insufficiency, while infants after the newborn stage need 1000–5000 IU of vitamin D per day for 8–12 weeks [15]. To treat vitamin D deficiency and prevent the hungry bone syndrome, which develops as a result of underlying hypocalcemia, vitamin D therapy is combined with appropriate calcium administration. To attain ideal serum levels, children with malabsorption syndromes and those taking anticonvulsants, glucocorticoids, antifungals, or antiretroviral drugs require longer and higher oral doses of vitamin D [16].

Dietary consumption and cutaneous production both contribute to the maintenance of vitamin D levels. Neonates and children have a greater potential to generate vitamin D from sunshine because they have a higher body surface area to volume ratio than adults [17]. A 10- to 15-minute period of direct sunshine exposure can generate 10,000 to 20,000 IU of vitamin D. It has been demonstrated that completely dressed infants exposed to sunlight for 2 hours each week can avoid severe vitamin D insufficiency [18]. However, it should be noted that exposure to direct sunshine is not recommended in infants younger than 6 months. Geographical latitude, degree of skin pigmentation, and the amount of skin exposed to sunlight are among the most critical variables that affect vitamin D synthesis from sunlight. Therefore, children with more skin pigmentation are at a higher risk of vitamin D deficiency. Compared to children with less skin pigmentation, these children need five to ten times more time in the sun for the same amount of 25(OH) vitamin D to be produced. Asian children need three times more sun exposure than white American children to maintain adequate vitamin D levels due to their darker complexion.

4. Vitamin D deficiency and neonatal and pediatric health outcomes

In addition to its well-defined classical functions related to calcium homeostasis and bone development, the relationship between vitamin D levels and health outcomes in infancy has also attracted exceptional attention in the scientific community. We begin our review of possible links between vitamin D status and health outcomes with pregnancy outcomes. Despite its critical importance, it is still unclear whether maternal vitamin D status plays a role in proper fetal and placental development and consequently in pregnancy outcomes. One recent review found no clear evidence to suggest that low vitamin D levels in early pregnancy are associated with adverse pregnancy outcomes, mainly preeclampsia, fetal growth restriction, preterm birth, and stillbirths [19]. In contrast, another review concluded that pregnant women with low 25(OHD) levels had an increased risk of gestational diabetes, preeclampsia, small for gestational age infants, and lower birth weight infants but no association with delivery by cesarean section [20]. Aghajafari et al. reviewed five randomized controlled trials and suggested a protective effect of vitamin D supplement during pregnancy on low birth weight (LBW) but no effect on preterm delivery [21].

The relationship between maternal vitamin D levels and particular child health outcomes, such as infections, has been investigated. Evidence suggests a relationship between maternal vitamin D levels and infants' predisposition to infection. One study found a significantly higher risk of respiratory infections (colds, cough, whooping cough, chest infection, and ear infection) by 3 months of age among infants with cord blood levels of 25(OH)D less than 25 nmol/L [22]. In contrast, a cohort study followed-up children at the age of 9 months and found that mothers in the top quartile of 25 (OH) D statuses in late pregnancy were significantly more likely to report their children having been diagnosed with pneumonia or bronchiolitis compared with those in the bottom quartile [23]. One recent study showed an association between early-onset neonatal sepsis and low maternal vitamin D levels in term infants [24]. Studies have also suggested that vitamin D pathways may be involved in the susceptibility to and outcome of Hepatitis B Virus infection acquired early in life [25].

The epidemiological evidence of the link between maternal vitamin D levels and infection is inconclusive, but vitamin D has a direct role in the production of antimicrobial peptides such as cathelicidin, which may help prevent infection during pregnancy and early childhood. This mechanism suggests a plausible biological basis for the relationship between maternal vitamin D levels and infection in infancy. Preventive measures in pregnant women aim to ensure that they have enough vitamin D, either from sunlight exposure or the right vitamin D supplements, and protect the newborn against possible adverse effects of vitamin D deficiency.

5. Vitamin D deficiency in neonatal and pediatric critical care

Accumulating evidence from various parts of the world points to a high prevalence of vitamin D deficiency among children admitted to neonatal and pediatric critical care units. Babies in the neonatal period are particularly vulnerable to a range of diseases requiring NICU hospitalization, and the study of vitamin D status in this group has attracted considerable interest in recent years. In a study by Bhimji et al. in Tanzania, about 80% of newborns admitted to the NICU of a tertiary care hospital showed vitamin D insufficiency [26]. A study from the U.S. discovered vitamin D inadequacy or insufficiency in 80% of preterm newborns with birth weights under 1500 g [27]. Another study from Australia found such conditions in 35.7% of preterm neonates admitted to NICU [28]. To determine the prevalence of vitamin D deficiency, Chacham et al. carried out an observational study on infants aged equal to or younger than 1 year at a tertiary care facility in Northern India. Neonates comprised 80% of the population under study and had a 79% prevalence of vitamin D insufficiency [29]. According to a recent study from Iran, 37% of neonates hospitalized in the NICU had vitamin D deficiency, while 58% had vitamin D insufficiency. Thus,

95% of neonates had abnormal vitamin D levels at admission [30]. Kim et al. studied vitamin D status in very-low-birth-weight neonates in Korea and reported a mean serum vitamin D level of 13.4 ± 9.3 ng/mL, with 79.8% of the subjects being vitamin D deficient. They found a higher prevalence of respiratory morbidities, such as bronchopulmonary dysplasia and respiratory distress syndrome in preterm neonates, that had severe vitamin D deficiency. Moreover, vitamin D deficiency was associated with a longer NICU stay [31].

Vitamin D status has also been studied beyond the neonatal period and in the pediatric critical care setting. Rey et al. studied vitamin D levels of critically ill children in PICU and healthy kids in Spain. They found a nearly twofold higher rate of vitamin D deficiency in PICU patients compared to healthy controls [32]. In a multicenter study of critically ill children across Canada by McNally et al., the mean 25(OH)D levels (43 nmol/L) were much lower than the 67–75 nmol/L mean levels observed among healthy Canadian and US children. In patients who needed catecholamine infusions, required mechanical breathing, and received more than 40 ml/kg of fluid resuscitation, 25(OH)D levels were lower. Vitamin D deficiency was also independently associated with a 2-day increase in PICU stays [33]. The authors hypothesized that, during critical illness, sudden reductions in vitamin D levels might be more physiologically relevant than chronic deficiency because compensatory mechanisms are affected by inflammation and multiorgan failure in this setting. Similarly, in their study of a tertiary care PICU in India, Sankar et al. found a vitamin D deficiency prevalence of 74%. Furthermore, vitamin D deficiency was associated with a longer duration of PICU stay in this study [34].

A prospective study by Madden et al. at Boston Children's Hospital's pediatric intensive care unit (PICU) examined vitamin D levels at the time of admission. The study did not include patients undergoing heart surgery. The 25(OH)D deficiency was significantly associated with poorer clinical outcomes. In patients with vitamin D deficiency, the scores of illness severity were higher, and they were more likely to need vasopressor treatment. However, no association was discovered between 25(OH)D levels and the time spent on mechanical ventilation, and ionized serum calcium levels were normal in almost all cases [35]. The authors suggested that these findings were secondary to vitamin D's function in immunological regulation, inflammation, and calcium homeostasis and not due to fluid shifts and hemodilution.

6. Vitamin D deficiency and critical care in congenital heart disease

Congenital heart disease (CHD) and its surgical treatment is one the conditions that necessitate ICU admission and critical care in neonates and pediatric patients. In recent years, a considerable amount of research has been dedicated to vitamin D status in children with cardiac disease and the relationship between vitamin D levels and treatment outcomes in this group. In the study by Rippel et al. on critically ill children, two-thirds of the patients were postoperative cardiac patients. The researchers found no association between 25(OH)D deficit and the requirement for mechanical ventilation, the need for vasoactive support, the length of hospital or ICU stays, the severity of disease scores, or mortality. However, the likelihood of being vitamin D deficient in patients with cardiac diseases was almost twice that of non-cardiac patients in this study (40% vs. 22%) [36].

In 2013, McNally et al. reported the findings of a prospective cohort study on the vitamin D status of 54 children with CHD who underwent open heart surgery at a mean age of 8.4 months and 4 children who had coarctation of the aorta and underwent closed heart surgery. Forty-two percent of patients had a preoperative 25OHD deficiency (<50 nM), with the mean value being 58.0 nM (SD, 22.4). Following surgery, there was a 40% reduction in mean 25OHD to 34.2 nM (SD, 14.5), with 86% of subjects having vitamin D deficiency. The open-heart surgery group's decline in vitamin D levels was more remarkable. Intraoperative measurements showed a sudden drop in vitamin D with the start of cardiopulmonary bypass. Additionally, the authors investigated the need for catecholamines and discovered an association between lower postoperative- but not preoperative- vitamin D levels and the need for catecholamines [37]. A study of the vitamin D levels in 20 children with CHD having open heart surgery was published in 2017 by Abou Zahr et al. Prior to surgery, 40% of the patients had 25OHD deficiency, with values <20 ng/mL. After cardiac bypass surgery, the investigators found that the mean vitamin D level had significantly decreased [38]. Dohain et al. examined the vitamin D status of 69 CHD children following open heart surgery in 2020. Of the patients, 34 (49.5%) had vitamin D deficiency before surgery, and 63 (91.3%) after surgery. They found a 42.03% reduction in 25(OH)D after surgery and noted an association between decreased postoperative vitamin D levels and a rise in inotropic support requirement. These findings suggest a link between unfavorable circumstances and the postoperative decline in vitamin D levels [39]. In 2021, Ye et al. evaluated the relationship between preoperative vitamin D deficiency and the maximum vasoactive-inotropic score 24 hours after surgery in 900 children with CHD. The median total serum 25(OH)D level before surgery was 24.0 ng/mL, and 32.6% of the patients had vitamin D deficiency (25(OH)D < 20 ng/mL). They discovered an association between low vitamin D levels and the need for more postoperative inotropic support 24 hours after cardiac surgery [40].

A number of studies have focused on vitamin D status in neonates with CHD in critical care settings. In 2013, Graham et al. studied the vitamin D status of 70 neonates with CHD having open heart surgery. Before the procedure, 84% (59/70) of the subjects had vitamin D insufficiency. There was no significant decline in vitamin D levels after the operation. However, higher inotropic support was required when postoperative 25(OH)D levels were lower [41]. In 2022, Mosayebi et al. studied changes in vitamin D status in neonates with CHD having heart surgery. The study included 33 open-heart surgery patients and 13 patients with closed-heart surgery. This wide coverage allowed researchers to compare vitamin D status in the open- and closed-heart surgeries. Before the procedure, 66.7% of the patients were vitamin D deficient. This number rose to 84.4% after surgery. After surgery, vitamin D levels declined significantly in patients who had open heart surgery but not those who had closed heart surgery. A significant association was found between the rate of postoperative decrease in vitamin D levels and an unfavorable outcome while preoperative vitamin D levels did not demonstrate a link with the outcome and did not predict the rate of postoperative vitamin D decline. The authors suggested a sharp decrease in vitamin D levels after surgery as an indicator of an unfavorable outcome [6].

Citing the evidence that links vitamin D deficiency to poor outcomes in neonate and infant CHD surgery, in 2020, McNally et al. conducted a dose evaluation feasibility study in preparation for a clinical trial of vitamin D supplementation in children with CHD undergoing corrective surgery. They reported that supplementation with a daily high dose of vitamin D before surgery improved vitamin D status at the time of pediatric ICU admission. The authors recommended modifying the trial protocol and giving vitamin D supplements to patients for at least 1 month before surgery or considering a loading dose [42].

7. Conclusion

With the gradual recognition of the crucial role of Vitamin D status in many physiological and pathological processes, vitamin D deficiency has been proposed as a possible modifiable risk factor for ICU outcomes. The discovery of the role of vitamin D in stress response in the immune, cardiovascular, and respiratory systems has lent support to this possibility. Moreover, numerous clinical studies have identified a high prevalence of vitamin D deficiency among ICU patients.

Neonates and pediatric patients constitute a particularly vulnerable group, and the vitamin D status in this group in critical illness has come under increasing scrutiny in recent years. The studies have predominantly shown a high prevalence of vitamin D deficiency in critically ill neonates and pediatric patients, pointing to a potential role for vitamin D status in critical illness in these patients. Vitamin D deficiency in CHD patients undergoing corrective surgical treatment has attracted particular attention, and a number of studies have focused on this topic. Overall, these studies report high prevalence rates of vitamin D deficiency in this group of neonates and pediatric patients. Moreover, several studies report significant associations between low vitamin D levels and unfavorable findings, such as increased requirements for vasoactive support and mechanical ventilation and prolonged ICU stays, in these patients.

An evaluation of available data suggests vitamin D deficiency as a modifiable risk factor in neonatal and pediatric critical illness, specifically in CHD patients undergoing surgical treatment. Clinical trials have been proposed to examine the potential beneficial effect of preoperational vitamin D supplementation on the outcome of heart surgery in this group. Studies on this topic are still in progress. However, for now, vitamin D supplementation should be considered in critically ill neonates in general and in those undergoing surgery for CHD in particular. Such supplementation aims to maintain serum/plasma 25(OH)D concentrations at safe levels over the threshold of vitamin D deficiency.

Conflict of interest

The author declares no conflict of interest.

Author details

Pedram Ghahremani
Medical School, Zanjan University of Medical Sciences, Zanjan, Iran

*Address all correspondence to: gharamanip@gmail.com

References

[1] Christakos S, Dhawan P, Verstuyf A, Verlinden L, Carmeliet G. Vitamin D: Metabolism, molecular mechanism of action, and pleiotropic effects. Physiological Reviews. 2016;**96**(1):365-408. DOI: 10.1152/physrev.00014.2015

[2] Caprio M, Infante M, Calanchini M, Mammi C, Fabbri A. Vitamin D: Not just the bone. Evidence for beneficial pleiotropic extraskeletal effects. Eating and Weight Disorders. 2017;**22**(1):27-41. DOI: 10.1007/s40519-016-0312-6

[3] Lee P, Nair P, Eisman JA, Center JR. Vitamin D deficiency in the intensive care unit: An invisible accomplice to morbidity and mortality? Intensive Care Medicine. 2009;**35**(12):2028-2032. DOI: 10.1007/s00134-009-1642-x

[4] Han JE, Jones JL, Tangpricha V, Brown MA, Brown LAS, Hao L, et al. High dose vitamin D Administration in Ventilated Intensive Care Unit Patients: A pilot double blind randomized controlled trial. Journal of Clinical & Translational Endocrinology. 2016;**4**:59-65. DOI: 10.1016/j.jcte.2016.04.004

[5] McNally JD, Nama N, O'Hearn K, Sampson M, Amrein K, Iliriani K, et al. Vitamin D deficiency in critically ill children: A systematic review and meta-analysis. Critical Care. 2017;**21**(1):287. DOI: 10.1186/s13054-017-1875-y

[6] Mosayebi Z, Mirzaaghayan M, Zeinaloo A, Gharib B, Ghahremani P. Vitamin D status in neonates with congenital heart disease: Its changes after heart surgery and relationship to the outcome. Journal of Advances in Medical and Biomedical Research. 2022;**30**(138):17-23. DOI: 10.30699/jambs.30.e53711

[7] Ross AC, Taylor CL, Yaktine AL, Del Valle HB, editors. Institute of Medicine (US) Committee to Review Dietary Reference Intakes for Vitamin D and Calcium. Dietary Reference Intakes for Calcium and Vitamin D. Washington, DC: National Academies Press; 2011. Available from: http://www.ncbi.nlm.nih.gov/books/NBK56070/ [Accessed: August 13, 2022]

[8] Holick MF, Binkley NC, Bischoff-Ferrari HA, Gordon CM, Hanley DA, Heaney RP, et al. Evaluation, treatment, and prevention of vitamin D deficiency: An Endocrine Society clinical practice guideline. The Journal of Clinical Endocrinology and Metabolism. 2011;**96**(7):1911-1930. DOI: 10.1210/jc.2011-0385

[9] Munns CF, Shaw N, Kiely M, Specker BL, Thacher TD, Ozono K, et al. Global consensus recommendations on prevention and Management of Nutritional Rickets. The Journal of Clinical Endocrinology and Metabolism. 2016;**101**(2):394-415. DOI: 10.1210/jc.2015-2175

[10] Scientific Advisory Committee on Nutrition. Report on Vitamin D and Health. 2016. Available from: https://www.gov.uk/government/publications/sacn-vitamin-d-and-health-report. [Accessed: August 13, 2022]

[11] Mithal A, Wahl DA, Bonjour JP, Burckhardt P, Dawson-Hughes B, Eisman JA, et al. Global vitamin D status and determinants of hypovitaminosis D. Osteoporosis International. 2009;**20**(11):1807-1820. DOI: 10.1007/s00198-009-0954-6

[12] Saraf R, Morton SMB, Camargo CA, Grant CC. Global summary of maternal

and newborn vitamin D status - a systematic review. Maternal & Child Nutrition. 2016;**12**(4):647-668. DOI: 10.1111/mcn.12210

[13] Karras SN, Shah I, Petroczi A, Goulis DG, Bili H, Papadopoulou F, et al. An observational study reveals that neonatal vitamin D is primarily determined by maternal contributions: Implications of a new assay on the roles of vitamin D forms. Nutrition Journal. 2013;**12**:77. DOI: 10.1186/1475-2891-12-77

[14] Balasubramanian S, Dhanalakshmi K, Amperayani S. Vitamin D deficiency in childhood-a review of current guidelines on diagnosis and management. Indian Pediatrics. 2013;**50**(7):669-675. DOI: 10.1007/s13312-013-0200-3

[15] Joiner TA, Foster C, Shope T. The many faces of vitamin D deficiency rickets. Pediatrics in Review. 2000;**21**(9):296-302. DOI: 10.1542/pir.21-9-296

[16] Misra M, Pacaud D, Petryk A, Collett-Solberg PF, Kappy M. Drug and therapeutics Committee of the Lawson Wilkins Pediatric Endocrine Society. Vitamin D deficiency in children and its management: Review of current knowledge and recommendations. Pediatrics. 2008;**122**(2):398-417. DOI: 10.1542/peds.2007-1894

[17] Abrams SA, Committee on Nutrition. Calcium and vitamin d requirements of enterally fed preterm infants. Pediatrics. 2013;**131**(5):e1676-e1683. DOI: 10.1542/peds.2013-0420

[18] Casey CF, Slawson DC, Neal LR. Vitamin D supplementation in infants, children, and adolescents. American Family Physician. 2010;**81**(6):745-748

[19] Mazahery H, Camargo CA, Conlon C, Beck KL, Kruger MC, von Hurst PR. Vitamin D and autism Spectrum disorder: A literature review. Nutrients. 2016;**8**(4):236. DOI: 10.3390/nu8040236

[20] Nassar N, Halligan GH, Roberts CL, Morris JM, Ashton AW. Systematic review of first-trimester vitamin D normative levels and outcomes of pregnancy. American Journal of Obstetrics and Gynecology. 2011;**205**(3):208.e1-208.e7. DOI: 10.1016/j.ajog.2011.03.058

[21] Aghajafari F, Nagulesapillai T, Ronksley PE, Tough SC, O'Beirne M, Rabi DM. Association between maternal serum 25-hydroxyvitamin D level and pregnancy and neonatal outcomes: Systematic review and meta-analysis of observational studies. BMJ. 2013;**346**:f1169. DOI: 10.1136/bmj.f1169

[22] Camargo CA, Ingham T, Wickens K, Thadhani R, Silvers KM, Epton MJ, et al. Cord-blood 25-hydroxyvitamin D levels and risk of respiratory infection, wheezing, and asthma. Pediatrics. 2011;**127**(1):e180-e187. DOI: 10.1542/peds.2010-0442

[23] Gale CR, Robinson SM, Harvey NC, Javaid MK, Jiang B, Martyn CN, et al. Maternal vitamin D status during pregnancy and child outcomes. European Journal of Clinical Nutrition. 2008;**62**(1):68-77. DOI: 10.1038/sj.ejcn.1602680

[24] Cetinkaya M, Cekmez F, Buyukkale G, Erener-Ercan T, Demir F, Tunc T, et al. Lower vitamin D levels are associated with increased risk of early-onset neonatal sepsis in term infants. Journal of Perinatology. 2015;**35**(1):39-45. DOI: 10.1038/jp.2014.146

[25] Chatzidaki V, Choumerianou D, Dimitriou H, Kouroumalis E, Galanakis E. Genetic variants associated with susceptibility to mother-to-child transmission of hepatitis B virus. European Journal of Gastroenterology & Hepatology. 2012;**24**(10):1185-1190. DOI: 10.1097/MEG.0b013e328356440f

[26] Bhimji KM, Naburi H, Aboud S, Manji K. Vitamin D status and associated factors in neonates in a resource constrained setting. International Journal Of Pediatrics. 2018;**2018**:9614975. DOI: 10.1155/2018/9614975

[27] Munshi UK, Graziano PD, Meunier K, Ludke J, Rios A. Serum 25 Hydroxy vitamin D levels in very low birth weight infants receiving Oral vitamin D supplementation. Journal of Pediatric Gastroenterology and Nutrition. 2018;**66**(4):676-679. DOI: 10.1097/MPG.0000000000001831

[28] Panda M, McIntosh J, Chaudhari T, Kent AL. Do maternal vitamin D levels influence vitamin D levels in preterm neonates? International Journal Of Pediatrics. 2019;**2019**:8613414. DOI: 10.1155/2019/8613414

[29] Chacham S, Rajput S, Gurnurkar S, Mirza A, Saxena V, Dakshinamurthy S, et al. Prevalence of vitamin D deficiency among infants in northern India: A hospital based prospective study. Cureus. 2020;**12**(11):e11353. DOI: 10.7759/cureus.11353

[30] Mosayebi Z, Sagheb S, Mirzendedel M, Movahedian AH. Serum vitamin D deficiency in NICU hospitalized neonates and its association with neonatal outcomes. Journal of Family & Reproductive Health. 2021;**15**(2):99-105. DOI: 10.18502/jfrh.v15i2.6450

[31] Kim I, Kim SS, Song JI, Yoon SH, Park GY, Lee YW. Association between vitamin D level at birth and respiratory morbidities in very-low-birth-weight infants. Korean Journal of Pediatrics. 2019;**62**(5):166-172. DOI: 10.3345/kjp.2018.06632

[32] Rey C, Sánchez-Arango D, López-Herce J, Martínez-Camblor P, García-Hernández I, Prieto B, et al. Vitamin D deficiency at pediatric intensive care admission. Jornal de Pediatria. 2014;**90**(2):135-142. DOI: 10.1016/j.jped.2013.08.004

[33] McNally JD, Menon K, Chakraborty P, Fisher L, Williams KA, Al-Dirbashi OY, et al. The association of vitamin D status with pediatric critical illness. Pediatrics. 2012;**130**(3):429-436. DOI: 10.1542/peds.2011-3059

[34] Sankar J, Lotha W, Ismail J, Anubhuti C, Meena RS, Sankar MJ. Vitamin D deficiency and length of pediatric intensive care unit stay: A prospective observational study. Annals of Intensive Care. 2016;**6**(1):3. DOI: 10.1186/s13613-015-0102-8

[35] Madden K, Feldman HA, Smith EM, Gordon CM, Keisling SM, Sullivan RM, et al. Vitamin D deficiency in critically ill children. Pediatrics. 2012;**130**(3):421-428. DOI: 10.1542/peds.2011-3328

[36] Rippel C, South M, Butt WW, Shekerdemian LS. Vitamin D status in critically ill children. Intensive Care Medicine. 2012;**38**(12):2055-2062. DOI: 10.1007/s00134-012-2718-6

[37] McNally JD, Menon K, Chakraborty P, Fisher L, Williams KA, Al-Dirbashi OY, et al. Impact of anesthesia and surgery for congenital heart disease on the vitamin d status of infants and children: A prospective

longitudinal study. Anesthesiology. 2013;**119**(1):71-80. DOI: 10.1097/ALN.0b013e31828ce817

[38] Abou Zahr R, Faustino EVS, Carpenter T, Kirshbom P, Hall EK, Fahey JT, et al. Vitamin D status after cardiopulmonary bypass in children with congenital heart disease. Journal of Intensive Care Medicine. 2017;**32**(8):508-513. DOI: 10.1177/0885066616652077

[39] Dohain AM, Almogati J, Al-Radi OO, Elassal AA, Zaher ZF, Fatani TH, et al. Serum vitamin D status following pediatric cardiac surgery and association with clinical outcome. European Journal of Pediatrics. 2020;**179**(4):635-643. DOI: 10.1007/s00431-019-03538-x

[40] Ye X, Dong S, Deng Y, Jiang C, Kong Y, Tang L, et al. Preoperative vitamin D deficiency is associated with higher vasoactive-inotropic scores following Pediatric cardiac surgery in Chinese children. Frontiers in Pediatrics. 2021;**9**:671289. DOI: 10.3389/fped.2021.671289

[41] Graham EM, Taylor SN, Zyblewski SC, Wolf B, Bradley SM, Hollis BW, et al. Vitamin D status in neonates undergoing cardiac operations: Relationship to cardiopulmonary bypass and association with outcomes. The Journal of Pediatrics. 2013;**162**(4):823-826. DOI: 10.1016/j.jpeds.2012.10.013

[42] McNally JD, O'Hearn K, Fergusson DA, Lougheed J, Doherty DR, Maharajh G, et al. Prevention of post-cardiac surgery vitamin D deficiency in children with congenital heart disease: A pilot feasibility dose evaluation randomized controlled trial. Pilot and feasibility studies. 2020;**6**:159. DOI: 10.1186/s40814-020-00700-3

Chapter 4

Vitamin D Deficiency in Childhood Obesity: Behavioral Factors or Altered Metabolism?

Teodoro Durá-Travé and Fidel Gallinas-Victoriano

Abstract

Obesity childhood is related to vitamin D deficiency, but the mechanisms for this association still remain questionable. We hypothesized that behavioral factors would be decisive in reducing the body content of vitamin D in patients with obesity. A cross-sectional clinical and analytical study (calcium, phosphorus, calcidiol, and parathyroid hormone) was carried out in a group of 377 patients with obesity (BMI-DS >2.0), 348 patients with severe obesity (BMI-DS >3.0), and 411 healthy children. The place of residence was categorized as urban or rural. Vitamin D status was defined according to the US Endocrine Society criteria. The prevalence of vitamin D deficiency was significantly higher ($p < 0.001$) in severe obesity (48.6%) and obesity groups (36.1%) than in the control group (12.5%). Vitamin D deficiency was more frequent in severe obesity and obesity groups living in urban areas than in those living in rural areas (not in the control group). The patients with obesity living in urban residence did not present significant seasonal variations in vitamin D deficiency throughout the year in contrast to those patients with obesity living in rural residence. These findings suggest that the most probable mechanism for vitamin D deficiency in children and adolescents with obesity, rather than altered metabolic, is the behavioral factors (sedentary lifestyle and lack of adequate sunlight exposure).

Keywords: adolescents, children, calcidiol, obesity, parathyroid hormone, rural areas, urban areas, vitamin D

1. Introduction

Vitamin D is currently assigned a pleiotropic profile [1–3]. In point of fact, basically every human tissue and cell contains vitamin D receptors, and its biological effects are categorized as skeletal (bone metabolism and calcium homeostasis) and extra-skeletal (hypovitaminosis D appears to be involved in autoimmune diseases, infections, neuropsychiatric disorders, cardiovascular risk, prostate and breast cancer, etc.), a circumstance that justifies the interest in monitoring its body content.

Furthermore, the prevalence of childhood obesity has gradually increased in the course of the last decades, establishing as the most relevant nutritional disorder in our environment [4–6]. Even though obesity is considered as a multifactorial disorder, the celerity of its increase in prevalence is related essentially to behavioral factors: scarce healthy nutrition habits as well as a sedentary lifestyle conditioned, in large part, by new technologies (screen time, including television viewing, use of computers and video games) [7, 8].

Several studies have demonstrated that obesity childhood is related to vitamin D deficiency [9–12]. The main source of vitamin D is the exposure to natural sunlight (cutaneous synthesis through ultraviolet B radiation) and, therefore, the higher prevalence of vitamin D deficiency in children and adolescents with obesity could be secondary to a more sedentary lifestyle (less mobility and participation in outdoor activities) and, consequently, a lack of adequate sun exposure. However, many explanations have been proposed for this association, but, interestingly, they hardly introduce theoretical mechanisms that imply limited sun exposure: storage or sequestration in adipose tissue, volumetric dilution, impaired hepatic 25-hydroxylation, etc. [3, 13–16].

The main causes of vitamin D deficiency are generally ascribed either to some physical agent that obstructs solar radiation (clothing, sunscreen, etc.) or to geographical characteristics, such as latitude and season of the year, cloudy weather, altitude, etc. [2, 17]. In fact, recent studies using an objective and accurate method for ultraviolet radiation monitoring in children and adolescents have revealed that rural residents receive higher levels of ultraviolet radiation exposure than urban residents do [18, 19].

This study aims to compare vitamin D status between children and adolescents with obesity living in an urban area and in a rural area in Navarra, Spain (latitude between 43°16″42 and 41°55″22 North). We hypothesized that behavioral factors (outdoor activities and sun exposure) would be decisive in reducing the body content of vitamin D in patients with obesity.

2. Methods

2.1 Participants

We conducted a cross-sectional study in a group of 377 patients, aged 6.50–15.7 years, previously diagnosed with obesity (obesity group, BMI-DS >2.0, 97th percentile) and 348 patients, aged 7.4–15.3 years, diagnosed with severe obesity (severe obesity group, BMI-DS >3.0, 99th percentile). The participants were assessed in a clinical evaluation in the Pediatric Endocrinology Unit in this hospital in January 2014–December 2021. Clinical features (sex, age, season of study visit, place of residence, and BMI) and blood testing data (calcium, phosphorus, calcidiol, and PTH) were collected. Tanner's classification was used for the assessment of pubertal staging, and the individuals were subsequently classified in school subgroup (Tanner stage I) and adolescent subgroup (Tanner stages II–V). Additional classification based on the place of residence (population or the city, higher or less than 10,000 inhabitants, respectively) was made, establishing urban or rural subgroups.

These features and measurements (clinical examination and blood testing) were estimated in a control group made up of 411 healthy children, aged 7.1–14.9 years, with BMI-DS in a range of −1.0 (15th percentile) to +1.0 (85th percentile).

Every participant in the study was Caucasian and lived in Navarra, Spain. Clinical records were surveyed in order to exclude any condition that could affect bone health, or any chronic pathology that could affect growth, body composition, food ingestion, or physical activity, or the previous intake of any medication (antiepileptic drugs or glucocorticoids), vitamin D, or calcium supplements.

2.2 Clinical examination

A previously published standardized protocol was applied for the anthropometric measurements [20]: participants were placed in underwear and barefoot, and we used an Año-Sayol scale (reading interval 0–120 kg and a precision of 100 g) for weight measures, and a Holtain wall stadiometer (reading interval 60–210 cm, precision 0.1 cm) for height measures.

The program Aplicación Nutricional, from the Spanish Society of pediatric gastroenterology, hepatology, and nutrition (Sociedad Española de Gastroenterología, Hepatología y Nutrición Pediátrica, available at http://www.gastroinf.es/nutritional/) was used to calculate the standard deviation (DS) values for the BMI. The graphic charts from the study of Ferrández et al. (Centro Andrea Prader, Zaragoza 2002) were used as the reference pattern [21].

2.3 Blood testing

The blood sample for biochemical determinations (calcium, phosphorus, 25(0H) D, and PTH) was obtained in basal fasting conditions (between 8:00 h and 9:00 h after an overnight fast).

The medical device used for the determination of calcium and phosphorous plasma levels was a COBAS 8000 analyzer (Roche Diagnostic, Mannheim, Germany). The determination of calcidiol levels was made with a high-specific chemiluminiscence-immunassay (LIAISON Assay, Diasorin, Dietzenbach, Germany), and the determination of PTH levels using a highly specific solid-phase, two-site chemiluminescent enzyme-labeled immunometric assay in an Immulite analyzer (DPC Biermann, Bad Nauheim, Germany).

The criteria of the United States Endocrine Society [22, 23] were applied to distribute individuals according to vitamin D plasma levels. In this way, a determination of calcidiol plasma level below 20 ng/ml (<50 nmol/L) was considered vitamin D deficiency, calcidiol plasma levels between 20 and 29 ng/ml (50–75 nmol/L), vitamin D insufficiency, and concentrations equal to or higher than 30 ng/ml (> 75 nmol/L) vitamin D sufficiency.

2.4 Statistical analysis

Tables show the results as percentages (%) and means (M) with corresponding standard deviations (SD). The program Statistical Packages for the Social Sciences version 20.0 (Chicago, IL, USA) was used to perform the statistical analysis (descriptive statistics, Student's t-test, analysis of variance, $\chi 2$ test, and Pearson's correlation). Statistical significance was assumed when P value was <0.05.

Parents and/or legal guardians were informed and provided consent for the participation in this study in all cases. This study was approved by the Ethics Committee for Human Investigation at our institution (in accordance with the ethical standards laid down in the 1964 Declaration of Hensinki and later amendments).

3. Results

Table 1 shows and compares the distribution of demographic features in the severe obesity, obesity, and control groups. No significant differences were found in the

Variables	Control group (n = 411)	Obesity group (n = 377)	Severe obesity (n = 348)	p values*
Sex				
Boys	188 (45.7%)	176 (46.7%)	172 (47.8%)	0.754
Girls	223 (54.3%)	201 (53.3%)	188 (52.2%)	
Age Group				
Childhood	175 (42.6%)	118 (31.3%)	111 (30.8%)	0.420
Adolescent	236 (57.4%)	259 (68.7%)	249 (69.2%)	
Season of study visit				
Winter	128 (31.1%)	99 (26.3%)	83 (23.1%)	0.364
Spring	87 (21.2%)	80 (21.2%)	84 (23.3%)	
Summer	54 (13.1%)	69 (18.3%)	75 (20.8%)	
Autumn	142 (34.5%)	129 (34.2%)	118 (32.8%)	
Residence				
Urban	229 (55.7%)	181 (48.7%)	173 (48.1%)	0.553
Rural	182 (44.3%)	191 (51.3%)	187 (51.9%)	

*Chi2.

Table 1.
Distribution of geographic/demographic features in severe obesity, obesity, and control groups.

distribution in relation to sex, age group, season of blood sample collection, and place of residence.

The mean values for age in the severe obesity, obesity, and control groups were 11.4 ± 2.9, 11.3 ± 2.7, and 11.1 ± 2.5 years, respectively, and there were no significant differences (p = 0.650) in age between the different groups. Obviously, the mean values for BMI-SD were significantly higher (p < 0.001) in the severe obesity (4.27 ± 1.28) and obesity groups (2.48 ± 0.28) with respect to control group (0.24 ± 0.23).

Figure 1 depicts and compares the prevalence of vitamin D status in the control, obesity, and severe obesity groups. The prevalence of vitamin D deficiency was significantly higher (Chi2: 159.8, p < 0.001) in severe obesity (48.6%) and obesity groups (36.1%%) than in the control group (12.5%). That is, only 12.2% and 16.2% of

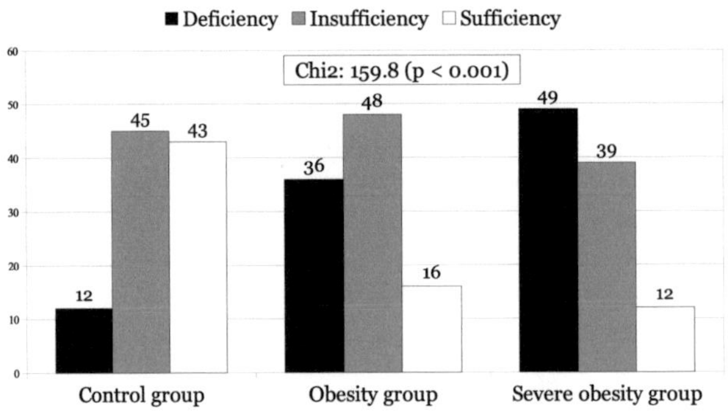

Figure 1.
Prevalence of vitamin D status in control, obesity, and severe obesity groups.

Groups	Deficiency	Insufficiency	Sufficiency	p value*
Control:				
Calcium (mg/dL)	10.0 ± 0.3	9.9 ± 0.3	9.9 ± 0.3	0.714
Phosphorus (mg/dL)	4.6 ± 0.5	4.5 ± 0.5	4.6 ± 0.6	0.670
PTH (pg/mL)	35.5 ± 14.9**	31.8 ± 13.9**	28.5 ± 12.5**	0.001
Calcidiol (ng/mL)	15.8 ± 2.6	24.5 ± 2.6	35.2 ± 4.5	0.001
Obesity:				
Calcium (mg/dL)	9.8 ± 0.3	9.9 ± 0.6	9.9 ± 0.3	0.416
Phosphorus (mg/dL)	4.6 ± 0.5	4.5 ± 0.6	4.5 ± 0.5	0.311
PTH (pg/mL)	55.5 ± 19.4**	46.1 ± 18.7**	40.6 ± 16.9**	0.001
Calcidiol (ng/mL)	14.2 ± 3.7	24.0 ± 2.7	35.3 ± 4.6	0.001
Severe obesity:				
Calcium (mg/dL)	9.7 ± 0.3 (173)	9.8 ± 0.3	9.9 ± 0.3	0.216
Phosphorus (mg/dL)	4.5 ± 0.6	4.5 ± 0.6	4.3 ± 0.5	0.689
PTH (pg/mL)	62.0 ± 21.3**	50.2 ± 18.4**	41.3 ± 15.3**	0.001
Calcidiol (ng/mL)	13.9 ± 3.8	24.0 ± 2.5	36.0 ± 5.3	0.001

**ANOVA.*
***ANOVA between groups (p < 0.001).*

Table 2.
Biochemical determinations according to vitamin D status in severe obesity, obesity, and control groups (M ± SD).

patients of the severe obesity and obesity groups showed levels of 25 (OH)D higher than 30 ng/mL, respectively, in contrast to 42.6% of the participants in the control group ($p < 0.01$).

Table 2 shows and compares the mean values for biochemical determinations in both groups in accordance to vitamin D status. There were not any significant differences in calcium and phosphorus levels between the different groups of vitamin D status, and obviously 25(OH)D levels were significantly lower ($p < 0.001$) in vitamin D insufficiency and deficiency individuals than in vitamin D sufficiency individuals in each group. PTH levels were significantly higher ($p < 0.001$) in the group with vitamin D insufficiency and deficiency than in vitamin D sufficiency within each group. In addition, there were not any significant differences in calcium, phosphorus, and 25(OH)D levels in each vitamin D status group between the different groups. However, PTH levels were significantly higher ($p < 0.001$) for each vitamin D status in the severe obesity and obesity groups with respect to the control group.

Figure 2 presents and compares the prevalence of vitamin D deficiency according to the seasons of the year between control, obesity, and severe obesity groups. In each group, the highest prevalence of vitamin D deficiency (Chi2: 65.01, $p < 0.001$) corresponded to winter (severe obesity group: 65.1%, obesity group: 40.4%, and control group: 19.5%), and they reached a minimum in the summer (severe obesity group: 26.7%, obesity group: 26.1%, and control group: 3.8%). The prevalence of vitamin D deficiency in the different seasons of the year was significantly higher ($p < 0.001$) in the severe obesity and obesity groups with respect to the control group.

Figure 3 exposes and compares the prevalence of vitamin D deficiency in relation to the place of residence between control, obesity, and severe obesity groups. In the

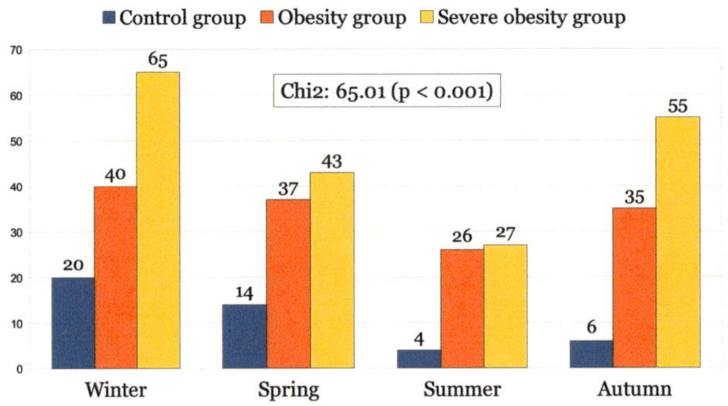

Figure 2.
Prevalence of vitamin D deficiency according to the seasons of the year in control, obesity, and severe obesity groups.

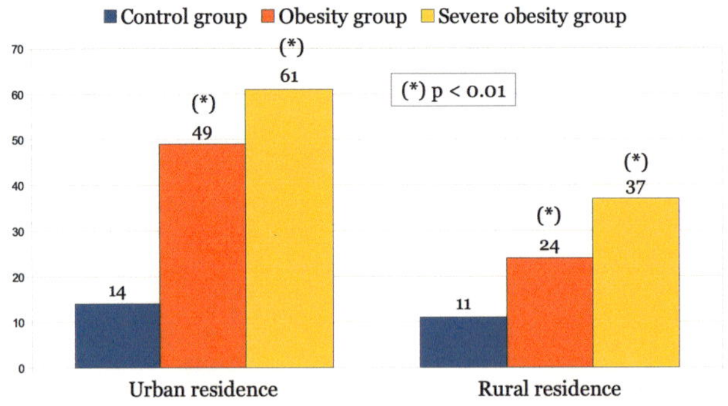

Figure 3.
Prevalence of vitamin D deficiency in relation to the place of residence in control, obesity, and severe obesity groups.

control group, there were no significant differences (p = 0.466) in vitamin D deficiency between urban (14.7%) and rural (10.5%) subgroups. As for the obesity group, vitamin D deficiency was significantly more frequent (p < 0.01) in the urban (48.6%) than in the rural subgroup (24.1%); additionally, in the severe obesity group, also vitamin D deficiency was significantly more frequent (p < 0.01) in the urban (61.3%) than in the rural subgroup (36.9%).

Figure 4 displays and compares the prevalence of vitamin D deficiency according to the seasons of the year between the individuals in the control, obesity, and severe obesity groups that lived in urban residence. In the control group, there were significant seasonal variations (Chi2: 38.1, p < 0.01) in vitamin D deficiency, which showed the lowest prevalence of vitamin D deficiency during the summer (7.1%) and the highest during the winter (25%). In contrast, there were no significant seasonal variations in the prevalence of vitamin D deficiency throughout the year in both the severe obesity and obesity groups. In fact, in severe obesity group, the prevalence of vitamin D deficiency during the summer was 51.6% and during the winter 67.6% (Chi2: 9.1, p = 0.170), and within the obesity group, vitamin D deficiency was 47% in the summer and 49.6% in the winter (Chi2: 9.2, p = 0.161).

Figure 5 shows and compares the prevalence of vitamin D deficiency according to the seasons of the year between the participants in the control, obesity, and severe obesity groups that lived in rural residence. All groups presented significant seasonal variations in vitamin D deficiency throughout the year. In each group, the lowest

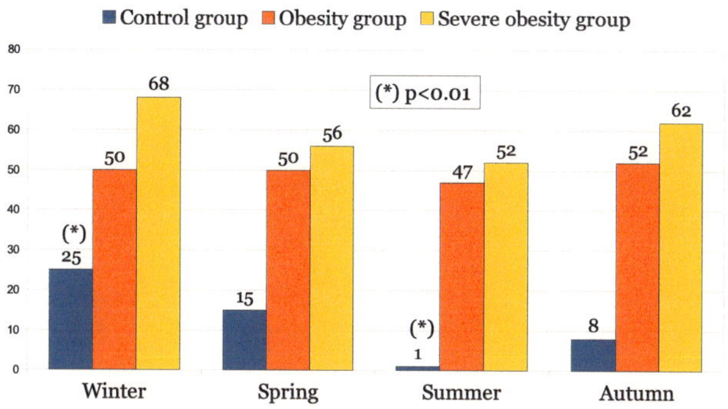

Figure 4.
Prevalence of vitamin D deficiency according to the seasons of the year in individuals in the control, obesity, and severe obesity groups that lived in urban residence.

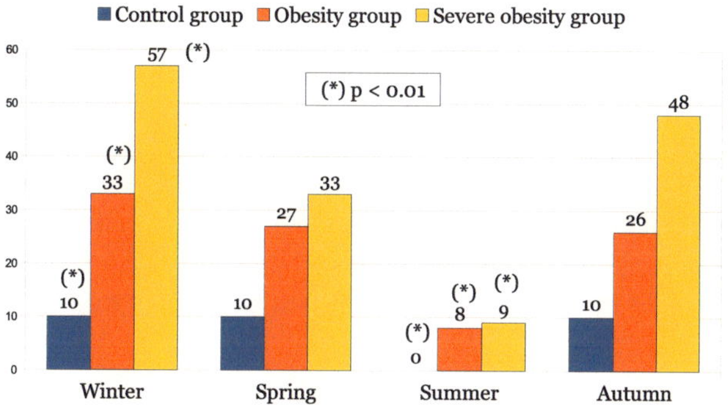

Figure 5.
Prevalence of vitamin D deficiency according to the seasons of the year in participants in control, obesity, and severe obesity groups that lived in rural residence.

prevalence of vitamin D deficiency corresponded to summer, and they reached a maximum in the winter. In severe obesity group, the prevalence of vitamin D deficiency during the summer was 9.1% and during the winter 57.1% (Chi2: 50,1, $p < 0.01$). In obesity group, vitamin D deficiency was 7.7% in the summer and 33.3% in the winter (Chi2: 21,9, $p < 0.01$). And, finally, in the control group, vitamin D deficiency was 0.0% during the summer and 10.4% during the winter (Chi2: 27,9, $p < 0.01$).

A negative correlation ($p < 0.01$) between calcidiol and PTH levels ($r = -0.375$) was detected. In addition, a positive correlation ($p < 0.01$) between PTH and BMI-SD ($r = 0.345$) and a negative correlation ($p < 0.01$) between calcidiol and BMI-SD ($r = -0.363$) were observed.

4. Discussion

This study verifies that vitamin D deficiency is a common condition in children and adolescents with obesity. Furthermore, our data suggest that this higher prevalence of vitamin D deficiency in these patients could be ascribed to inadequate sunlight exposure, since there was a weaker trend to vitamin D deficiency in those patients living in rural areas than in those living in urban areas.

Different geographic/demographic specificities, such as gender, age, season, or place of residence, have been described as factors associated with vitamin D deficiency [9–12, 24]; however, in this case, no significant differences were detected in the distribution of these characteristics among the participants included in this study (severe obesity, obesity, and control groups). This eventuality allows the comparison of the results obtained, avoiding confounding factors. The age range selected in the different groups of participants was due to the fact that they usually have enough autonomy to carry out their extracurricular and/or leisure activities at these ages.

The higher prevalence of vitamin D deficiency in obesity has been sustained by several studies [9–12], even though the potential mechanisms for this association still remain questionable. Nevertheless, at present, the most qualified hypotheses about the inverse relationship between vitamin D deficiency and obesity refer either to storage or sequestration of vitamin D in adipose tissue or volumetric dilution of vitamin D. Clinical studies have shown that obesity does not affect the cutaneous synthesis of vitamin D, but as it is a fat-soluble vitamin, it is accumulated and retained in the adipose tissue (storage site or sequestration hypothesis). Therefore, the greater the storage capacity of this vitamin in adipose tissue (severe obesity and obesity groups), the lower the serum levels of calcidiol [25, 26]. In fact, we found that calcidiol levels in the participants included in this study (severe obesity, obesity, and control groups) were inversely correlated with body mass index; this is an anthropometric measurement that has been frequently used in the diagnosis and follow-up of children and adolescents with obesity since it shows a good correlation with body fat content [27, 28]. A second probable mechanism of the inverse relationship between vitamin D deficiency and obesity could be a volumetric dilution; that is, vitamin D would be distributed in body compartments that increase with obesity (serum, fatty tissue, liver, etc.), thereby making serum levels lower [13, 15]. It has also been suggested that lower levels of calcidiol in obese patients could be due to impaired hepatic 25-hydroxylation related to nonalcoholic fatty liver disease, a condition that is common in obese adults but less frequent in childhood obesity [29]. However, none of the previously mentioned hypotheses would explain by itself, for example, the stronger trend to vitamin D deficiency in patients with obesity (severe obesity and obesity groups) living in urban areas than in those living in rural areas, as we identified in this study.

Vitamin D receptors are present in a large variety of tissues and cells in the body (muscle, heart, blood vessels, neurons, immune cells, breast, colon, prostate, etc.), and additionally, they have the capacity to produce calcitriol from circulating calcidiol. This fact supports the biological importance of sufficient calcidiol serum levels [1, 22]. Moreover, adipose tissue also expresses vitamin D receptors, and 1α-hydroxylase enzyme locally converts calcidiol to calcitriol (biological active form of vitamin D), and that process is not regulated by parathyroid hormone, in contrast with renal 1α-hydroxylase [30]. Additionally, some experimental data support that vitamin D could have an antiobesity effect by inhibiting adipogenesis during early adipocyte differentiation and independently of PTH. That is, vitamin D might be implicated in the pathogenesis of obesity, rather than being a consequence [3, 16]. These findings suggest, on one side, that adipose tissue could play a role in vitamin D metabolism rather than being a passive store of fat-soluble nutrients and, on the other side, that a bidirectional causal relationship between vitamin D deficiency and obesity cannot be excluded. However, several studies have shown no effect of vitamin D treatment on reducing body weight and/or body composition, suggesting that although vitamin D deficiency is associated with obesity, it is not bidirectional [31, 32].

In accordance with most authors [9, 10, 12, 24], we found a negative correlation between PTH and calcidiol levels, and this would be consistent with the physiological feedback mechanism of vitamin D on PTH secretion. But, interestingly, it is worth noting our finding that PTH levels were also significantly higher—independent of vitamin D status—in the patients with obesity (severe obesity and obesity groups) with respect to the control group. Many researchers have postulated that this elevation of PTH might increase calcium influx into adipocytes, which then leads to increased lipogenesis and potentially reduces catecholamine-induced lipolysis and, consequently, fosters fat storage [33, 34]. Additionally, several observational studies have shown that PTH levels in obesity are independent of vitamin D status, and it does not represent, as is commonly assumed, secondary hyperparathyroidism from hypovitaminosis D [35]. However, despite the above biological assumptions that obesity is related with vitamin D deficiency and elevated parathyroid hormone levels, the reason given for this association remains unexplained. In fact, some authors are currently questioning whether vitamin D deficiency is a consequence or cause of obesity [16], or whether the association between obesity and vitamin D deficiency is causality or casualty [3].

Obviously, unhealthy eating habits are related to childhood obesity, and this entails a lower intake of vitamin D. However, the main source of vitamin D is exposure to natural sunlight, while approximately 10% comes from natural dietary sources [1, 2]. Few foods naturally contain vitamin D (oily fish such as salmon, sardines, mackerel, and tuna, as well as shiitake mushrooms and eggs yolk) and, depending on the country, additional sources include fortified foods such as dairy products, orange juice, breakfast cereals, cookies, and butter or margarine [2, 36]. Therefore, even though diet seems to be probably an irrelevant factor in the acquisition of optimal levels of vitamin D, it could not be completely excluded.

Because geographical conditions affect body vitamin D levels, we cannot refer to a vitamin D status in a determined population without mentioning them. In our case, it should be noted that Navarre is a Spanish region located in the north of the Iberian peninsula with a population of 661,537 inhabitants (2021 census, National Institute of Statistics), 58.1% of whom live in urban areas and 41.9% in rural areas. Besides, it is characterized by a high frequency of precipitations and/or cloudiness and, especially, a high latitude (between 41°55″22 and 43°16″42 North). When the zenith angle of the sun is oblique, as occurs in the winter months in both hemispheres, type B ultraviolet radiation barely reaches the earth's surface above and below 40°N and 40°S latitude, causing a very low or absence cutaneous synthesis of vitamin D, even with prolonged sun exposure [17, 22, 23]. In compliance with several studies [9, 24, 37, 38], this is a potential explanation for the seasonal variations in the prevalence of vitamin D deficiency (maximum prevalence in the winter months and minimum in the summer months) that we found in the control group.

Recent studies using personal electronic ultraviolet radiation dosimeters have displayed higher ultraviolet radiation exposure in children and adolescents living in rural areas compared with those living in urban areas due to differences in types of activity. Children and adolescents living in rural areas spend more time after school and during weekdays practicing outdoors chores during peak ultraviolet radiation hours (10 am–4 pm), compared with those living in urban areas who spend more time participating in indoor sports and/or leisure activities and, therefore, reducing exposure to sunlight [18, 19]. These data allowed us to hypothesize a much simpler explanation for the relationship between obesity and vitamin D deficiency: behavioral factors (outdoor activities and sun exposure) would be determined in reduced body content of vitamin D in patients with obesity.

Indeed, we also found seasonal variations in the prevalence of vitamin D deficiency (maximum prevalence in the winter months and minimum in the summer months) in patients with obesity (obesity and severe obesity groups), although showing significantly lower values with respect to control group. That is, on the one hand, this would confirm that sunlight exposure has a large impact on vitamin D status also in patients with obesity and, on the other hand, we found a stronger trend to vitamin D deficiency in patients with obesity (obesity and severe obesity groups) living in urban areas than in those living in rural areas. No significant differences were observed in the prevalence of vitamin D deficiency in the control group in relation to place of residence (urban or rural). Nevertheless, the most remarkable finding of this study was that patients with obesity (obesity and severe obesity groups) living in urban residence did not present significant seasonal variations in vitamin D deficiency throughout the year in contrast to those patients with obesity (obesity and severe obesity) living in rural residence, who presented a maximum prevalence of vitamin D deficiency in the winter months and a minimum in the summer months. Therefore, these findings would support the hypothesis that the greater tendency to present vitamin D deficiency in obese children and adolescents would be related to a sedentary lifestyle and, consequently, to the lack of adequate sun exposure. Otherwise, the participants of the control group, who presumably did not have a sedentary lifestyle, showed no differences in vitamin D status in relation to the place of residence (rural or urban).

5. Conclusion

At present, and despite the hypotheses recounted above, vitamin D photobiology suggests that the most probable mechanism for vitamin D deficiency in children and adolescents with obesity, rather than altered metabolic (sequestration in adipose tissue, volumetric dilution, impaired hepatic 25-hydroylation, etc.), is the behavioral factors (reduced sunlight exposure), such as our findings outline. However, other mechanisms cannot be completely excluded, as they may contribute concurrently.

Funding

The authors received no financial support for the research, authorship, and/or publication of this article (none declared).

Conflict of interest

The authors declared no potential conflicts of interest with respect to the research, authorship, and/or publication of this study (none declared).

Author details

Teodoro Durá-Travé[1,2,3]* and Fidel Gallinas-Victoriano[2,3]

1 Department of Pediatrics, School of Medicine, University of Navarra, Pamplona, Spain

2 Department of Pediatrics, Navarra University Hospital, Pamplona, Spain

3 Navarrabiomed (Biomedical Research Center), Pamplona, Spain

*Address all correspondence to: tdura@unav.es

References

[1] Hossein-nezhad A, Holick MF. Vitamin D for health: A global perspective. Mayo Clinic Proceedings. 2013;**88**:720-755. DOI: 10.1016/j.mayocp.2013.05.011

[2] Palermo NE, Holick MF. Vitamin D, bone health, and other health benefits in pediatric patients. Journal of Pediatric Rehabilitation Medicine. 2014;**7**:179-192. DOI: 10.3233/PRM-140287

[3] Migliaccio S, Di Nisio A, Mele C, Scappaticcio L, Savastano S, Colao A, et al. Obesity and hypovitaminosis D: Causality or casuality? International Journal of Obesity. 2019;**9**:20-31. DOI: 10.1038/s41367-019-0010-8

[4] Spinelli A, Buoncristiano M, Kovacs VA, Yngve A, Spiroski I, Obreja G, et al. Prevalence of severe obesity among primary school children in 21 European countries. Obesity Facts. 2019;**12**:244-258. DOI: 10.1159/000500436

[5] Di Cesare M, Sorić M, Bovet P, Miranda JJ, Bhutta Z, Stevens GA, et al. The epidemiological burden of obesity in childhood: A worldwide epidemic requiring urgent action. BMC Medicine. 2019;**17**:212. DOI: 10.1186/s12916-019-1449-8

[6] Nittari G, Scuri S, Petrelli F, Pirillo I, di Luca NM, Grappasonni I. Fighting obesity in children from European World Health Organization member states. Epidemiological data, medical-social aspects, and prevention programs. La Clinica Terapeutica. 2019;**170**:e223-e230. DOI: 10.7417/CT.2019.2137

[7] Zhang G, Wu L, Zhou L, Lu W, Mao C. Television watching and risk of childhood obesity: A meta-analysis. European Journal of Public Health. 2016;**26**:13-18. DOI: 10.1093/eurpub/ckv213

[8] Robinson TN, Banda JA, Hale L, Lu AS, Fleming-Milici F, Calvert SL, et al. Screen media exposure and obesity in children and adolescents. Pediatrics. 2017;**140**(Suppl. 2):S97-S101. DOI: 10.1542/peds.2016-1758K

[9] Vierucci F, Del Pistoia M, Fanos M, Gori M, Carlone G, Erba P, et al. Vitamin D status and predictors of hypovitaminosis D in Italian children and adolescents: A cross-sectional study. European Journal of Pediatrics. 2013;**172**:1607-1617. DOI: 10.1007/s00431-013-2119-z

[10] Turer CB, Lin H, Flores G. Prevalence of vitamin D deficiency among overweight and obese US children. Pediatrics. 2013;**131**:e152-e161. DOI: 10.1542/peds.2012-1711

[11] Bellone S, Esposito S, Giglione E, Genoni G, Fiorito C, Petri A, et al. Vitamin D levels in a paediatric population of normal weight and obese subjects. Journal of Endocrinological Investigation. 2014;**37**:805-809. DOI: 10.1007/s40618-014-0108-3

[12] Durá-Travé T, Gallinas-Victoriano F, Chueca-Guindulain MJ, Berrade-Zubiri S. Prevalence of hypovitaminosis D and associated factors in obese Spanish children. Nutrition & Diabetes. 2017;**7**:e248. DOI: 10.1038/nutd.2016.50

[13] Drincic AT, Armas LA, Van Diest EE, Heaney RP. Volumetric dilution, rather than sequestration best explains the low vitamin D status of obesity. Obesity. 2012;**20**:1444-1448. DOI: 10.1038/oby.2011.404

[14] Vanlint S. Vitamin D and obesity. Nutrients. 2013;**5**:949-956. DOI: 10.3390/nu50030949

[15] Carrelli A, Bucovsky M, Horst R, Cremers S, Zhang C, Bessler M, et al. Vitamin D storage in adipose tissue of obese and normal weight women. Journal of Bone and Mineral Research. 2016;**32**:237-242. DOI: 10.1002/jbmr.2979

[16] Vranic L, Mikolasevic I, Milie S. Vitamin D deficiency: Consequence or cause of obesity. Medicina. 2019, 2019;**55**:541. DOI: 10.3390/medicina55090541

[17] Misra M, Pacaud D, Petryk A, Collett-Solberg PF, Kappy M. Vitamin D deficiency in children and its management: Review of current knowledge and recommendations. Pediatrics. 2008;**122**:398-417. DOI: 10.1542/peds.2007-1894

[18] Griffin TP, Wall D, Blake L, Griffin DG, Robinson S, Bell M, et al. Higher risk of vitamin D insufficiency/deficiency for rural than urban dwellers. The Journal of Steroid Biochemistry and Molecular Biology. 2020;**197**:105547. DOI: 10.1016/j.jsbmb.2019.105547

[19] Nagelhout ES, Lensink R, Zhu A, Parsons BG, Haaland B, Hashibe M, et al. Higher ultraviolet radiation exposure among rural-dwelling versus urban-dwelling adults and children: Implications for skin Cancer prevention. Journal of Community Health. 2021;**46**:147-155. DOI: 10.1007/s10900-020-00860-6

[20] Durá-Travé T et al. Assessment of body composition changes during a combined intervention for the treatment of childhood obesity. Nutrition. 2019;**59**:116-120. DOI: 10.1016/j.nut.2018.08.007 0899-9007

[21] Ferrandez A, Baguer L, Labarta JI, Labena C, Mayayo E, Puga B, et al. Estudio longitudinal de ninos españoles normales desde el nacimiento hasta la edad adulta. 1st ed. Zaragoza: Fundacion Andrea Prader; 2005

[22] Holick MF, Binkley NC, Bischoff-Ferrari HA, Gordon CM, Hanley DA, Heaney RP, et al. Evaluation, treatment, and prevention of vitamin D deficiency: An Endocrine Society clinical practice guideline. The Journal of Clinical Endocrinology and Metabolism. 2011;**96**:1911-1930. DOI: 10.1210/jc.2011-0385

[23] Holick MF, Binkley NC, Bischoff-Ferrari HA, Gordon CM, Hanley DA, Heaney RP, et al. Guidelines for preventing and treating vitamin D deficiency and insufficiency revisited. The Journal of Clinical Endocrinology and Metabolism. 2012;**97**:1153-1158. DOI: 10.1210/jc.2011-2601

[24] Weng FL, Shults J, Leonard MB, Stallings VA, Zemel BS. Risk factors for low serum 25-hydroxyvitamin D concentrations in otherwise healthy children and adolescents. The American Journal of Clinical Nutrition. 2007;**86**:150-158. DOI: 10.1093/ajcn/86.1. 150

[25] Worstman J, Matsuoka LY, Chen TC, Lu Z, Holick MF. Decreased bioavailability of vitamin D in obesity. The American Journal of Clinical Nutrition. 2000;**72**:690-693. DOI: 10.1093/ajcn/72.3.690

[26] Blum M, Dolnikowski G, Seyoum E, Harris SS, Booth SL, Peterson J, et al. Vitamin D3 in fat tissue. Endocrine. 2008;**33**:90-94. DOI: 10.1007/s12020-008-9051-4

[27] Martin-Calvo N, Moreno-Galarraga L, Martinez-Gonzalez MA. Association between body mass index, waist-to-height ratio and adiposity in children: A systematic review and meta-analysis.

Nutrients. 2016;**20**:8. DOI: 10.3390/nu8080512

[28] Styne DM, Arslanian SA, Connor EL, Farooqi IS, Murad MH, Silverstein JH, et al. Pediatric obesity-assessment, treatment, and prevention: An Endocrine Society clinical practice guideline. The Journal of Clinical Endocrinology and Metabolism. 2017;**102**:709-757. DOI: 10.1210/jc.2016-2573

[29] Targher G, Bertolini L, Scala L, Cigolini M, Zenari L, Falezza G, et al. Associations between serum 25-hydroxyvitamin D3 concentrations and liver histology in patients with non-alcoholic fatty liver disease. Nutrition, Metabolism, and Cardiovascular Diseases. 2007;**17**:517-524. DOI: 10.1016/j.numecd.2006.04.002

[30] Abbas MA. Physiological functions of vitamin D in adipose tissue. The Journal of Steroid Biochemistry and Molecular Biology. 2017;**165**:369-338. DOI: 10.1016/j.jsbmb.2016.08.004

[31] Wamberg L, Kampmann U, Stødkilde-Jørgensen H, Rejnmark L, Pedersen SB, Richelsen B. Effects of vitamin D supplementation on body fat accumulation, inflammation, and metabolic risk factors in obese adults with low vitamin D levels—Results from a randomized trial. European Journal of Internal Medicine. 2013;**24**:644-649. DOI: 10.1016/j.ejim.2013.03.005

[32] Mason C, Xiao L, Imayama I, Duggan C, Wang CY, Korde L, et al. VitaminD3 supplementation during weight loss: A double-blind randomized controlled trial. The American Journal of Clinical Nutrition. 2014;**99**:1015-1025. DOI: 10.3945/ajcn.113. 073734

[33] McCarty MF, Thomas CA. PTH excess may promote weight gain by impeding catecholamine-induced lipolysis—Implications for the impact of calcium, vitamin D, and alcohol on body weight. Medical Hypotheses. 2003;**61**:535-542. DOI: 10.1016/S0306-9877(03)00227-5

[34] Mai XM, Chen Y, Camargo CA. Cross-sectional and prospective cohort study of serum25- hydroxyvitamin D level and obesity in adults: The HUNT study. American Journal of Epidemiology. 2012;**175**:1029-1036. DOI: 10.1093/aje/kwr456

[35] Grethen E, McClintock R, Gupta CE, Jones RM, Cacucci BM, Diaz D, et al. Vitamin D and hyperparathyroidism in obesity. The Journal of Clinical Endocrinology and Metabolism. 2011;**96**:1320-1326. DOI: 10.1210/jc.2010-2202

[36] Holick MF. Vitamin D deficiency. The New England Journal of Medicine. 2007;**357**:266-281. DOI: 10.1056/NEJMra070553

[37] Karagüzel G, Dilber B, Can G, Ökten A, Değer O, Holick MF. Seasonal vitamin D status of healthy schoolchildren and predictors of low vitamin D status. Journal of Pediatric Gastroenterology and Nutrition. 2014;**58**:654-660. DOI: 10.1097/MPG.000 0000000000274

[38] Rovner AJ, O'Brien KO. Hypovitaminosis D among healthy children in the United States a review of the current evidence. Archives of Pediatrics & Adolescent Medicine. 2008;**162**:513-519. DOI: 10.1001/archpedi.162.6.513

Section 3

Vitamin D Status and COVID-19

Chapter 5

The Role of Vitamin D in the Restriction of the Progress and Severity of COVID-19 Infection

Alakesh Bharali, Bhargab Deka, Himangshu Sarma, Ashique Ahmed, Bedanta Bhattacharjee, Santa Sarma, Suman Kumar, Susankar Kushari and Rajlakshmi Devi

Abstract

SARS-CoV-2 has affected socio-economic activity in every country around the world since its outbreak began in 2019. 3.5 million people have died worldwide as of now, including 3.2 lakh in India. The cytokine storm significantly contributes to COVID mortality. To put it simply, the virus causes an uncontrolled release of cytokines, which results in severe inflammation, multi-organ failure, and death. Vitamin D was discovered to be a significant risk factor for cytokine storm in COVID patients. Numerous studies have demonstrated that those with deficient serum vitamin D levels have a significant mortality rate. The current understanding of the role of vitamin D in immune modulation in the innate and adaptive immune systems and how this may relate to COVID-19 is discussed in this article. Additionally, we evaluated the most recent clinical information about vitamin D deficiency, cytokine storm, and COVID-19 mortality.

Keywords: Covid-19, cytokine storm, immune regulation, pandemic, vitamin D

1. Introduction

As with SARS (severe acute respiratory syndrome) and MERS (Middle East respiratory syndrome), the rapid spread of SARS-CoV-2 (severe acute respiratory syndrome coronavirus 2) wrecked global health and economics. As of May 13, 2022, the SARS-CoV-2 outbreak has caused more than 519.9 million infections with over 6.3 million deaths [1] (https://www.worldometers.info/coronavirus). The SARS-CoV-2 infected or COVID-19 patients may develop various clinical manifestations,including severe acute pulmonary disease [2, 3], heart damage [4], kidney injury [4], hepatic dysfunction [4], pancreatic symptoms [5], gastrointestinal [6] and olfactory dysfunction [7]. However, the impact of COVID-19 may have long-lasting effects on the physiological systems. Nevertheless, these numbers are increasing at a frenetic pace [8]. In certain people, COVID-19 infection is accompanied by a "cytokine storm." The cytokine storm is a phenomenon in which pro-inflammatory cytokines are released at an abnormally rapid rate, allowing for the recruitment of additional immune cells to the injury site, resulting in organ damage. Numerous investigations evaluating COVID-19 patient cytokine profiles discovered that the cytokine storm is associated with significant lung injury and potentially multi-organ failure in COVID-19 patients [9].

Recent research has revealed that different people's bodies respond differently to the infection. Despite sharing the same race, culture, age, and sex, some people contract the virus while others do not exhibit the core symptoms/remain asymptomatic. According to studies, some people's immune systems handle COVID-19 better than others. COVID-19 is more prevalent in those with pre-existing health disorders such as cardiovascular disease, diabetes, respiratory disease, or hypertension, particularly those over 60. These comorbidities/factors impair the immune system, hence raising the severity of the disease [10]. Vitamin D deficiency and cytokine storm have been identified as risk factors; more studies have shown that vitamin D deficiency causes chronic illness and mortality in COVID-19 patients. Vitamin D supplementation has been proven to reduce the likelihood of a cytokine storm, although the molecular mechanism is unknown [11]. Additionally, vitamin D has been identified to dampen cytokine storms during the 1918–1919 viral influenza pandemic and prior coronavirus pandemics [12]. Thus, vitamin D's astounding quality draws our attention to authoring this review on its role in lowering the COVID-19-associated cytokine storm.

2. Pathology of SARS-CoV-2 infection

Chinese researchers utilized Cryo-electron microscopy to demonstrate how SARS-CoV-2 infects humans. The virus targets angiotensin-converting enzyme II (ACE II), located in human cells and tissues such as the lungs, heart, kidneys, and intestines [13]. According to Mark Fielder, a scientist at Kingston University, the virus appears to target two types of lung cells: goblet cells, which coat the respiratory

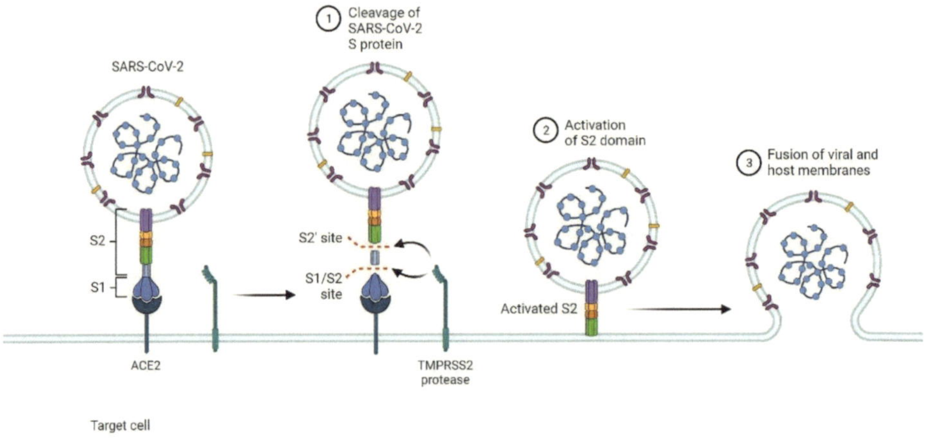

Figure 1.
Steps in the infection of SARS-CoV-2. The spike protein on the surface of the new coronavirus serves as the principal mechanism by which the virus enters the cell. A protein termed transmembrane protease serine-2 (TMPRSS2) cleaves spike protein into two subunits, S1 and S2. It is a cell surface protein that epithelial cells express in various organs, including those of the aerodigestive tract. TMPRSS2 aids the virus in infecting the host cell.

tract with mucus, and ciliated cells, filtering out debris [14]. Two receptor-binding domains (RBD) are present on the viral spike protein, one pointing downward and the other facing upward. As a result, the virus can bind to and infect human cells. The virus interacts with upward-facing RBD via spike proteins. Once inside the cell, the virus breaks down its protein coat and releases its Ribonucleic acid (RNA) payload, according to the British Society for Immunology. It replicates itself by taking control of the endoplasmic reticulum and the host cell structure. As a result of hijacking, the Golgi bodies of hijacked cells become encased in a protein shell that caps viral RNA and proteins, creating new viruses that exit the infected cell via the membrane. Coronavirus infection disturbs the host cell, according to a study published in Frontiers in Microbiology. Apoptosis, or cell death, occurs when an infection overwhelms the host cell's ability to sustain homeostasis (**Figure 1**) [15].

3. Role of vitamin D in immune regulation

Vitamin D is produced by the skin when it is exposed to sunshine. To begin, cholesterol is converted to 7-dehydrocholesterol by an enzyme called DHCR7 (7-Dehydrocholesterol Reductase). It is then converted to pre-vitamin D by the sun's ultraviolet B (UV-B) light, which spontaneously turns into vitamin D at body temperature. Although nature contains five different forms of vitamin D, only vitamin D_3, also known as cholecalciferol, is synthesized by the human body [16]. Cholecalciferol is an inactive form of vitamin D that the body must activate. It is converted to calcidiol in the liver by the enzyme vitamin D 25-hydroxylase. Calcidiol is then converted in the kidney to calcitriol (1,25-dihydroxycholecalciferol), the active form of vitamin D [17]. It is primarily regulated by parathyroid hormone, blood calcium/phosphorus levels, and calcitriol levels. Its active form aided in calcium absorption in the small intestine and was historically used to treat osteoporosis and rickets.

Additionally, it assists in the form of calcium deposits in the bones. Vitamin D has been recognized as a critical immune modulator and is also implicated in various

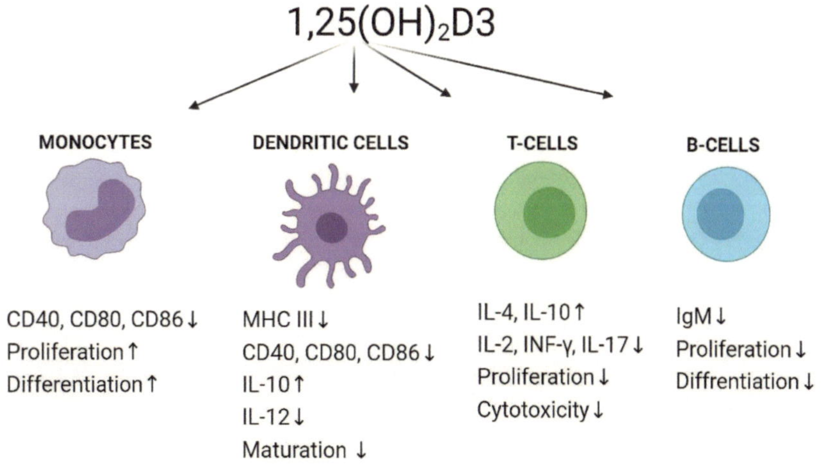

Figure 2.
Role of vitamin D in immune regulation. It down-regulates pro-inflammatory substances and up-regulates anti-inflammatory substances.

diseases, including autoimmune illnesses. The significant 1969 discovery that vitamin D, like other steroid vitamins such as vitamin A, may act on various cell types via activation of the vitamin D receptor (VDR) revealed a new window into understanding its role in health and disease [18]. As a result, vitamin D has been discovered as an immunomodulator, and VDR is detected in nearly all immune cells. Vitamin D increases IL-10 and T regulatory cells while decreasing T helper cytokines like IL-2, IL-32, Interferon-gamma (IFN-γ), and IL-17 [19]. This results in a change in damage-associated molecular pattern signaling, which is linked to numerous disorders (**Figure 2**).

4. Vitamin D as a regulator of the innate immune system

Vitamin D can augment the innate immune system's antimicrobial defenses. This role was discovered 30 years ago. Previous research has revealed that our immune cells, notably macrophages and monocytes, may synthesize vitamin D on their own and boost the development of an antimicrobial protein called cathelicidin, enhancing the intracellular clearance of *Mycobacterium tuberculosis* [20]. In epithelia, similar processes that contribute to barrier function have been discovered. Vitamin D receptor complexes enhance the transcription of cathelicidin by binding to vitamin D response sites in the cathelicidin gene promoter [21]. Cathelicidin is a significant antimicrobial protein with additional functions, including chemotaxis stimulation, generation of pro-inflammatory cytokines, recruitment of T-lymphocytes to the site of infection, and clearance of respiratory pathogens by apoptosis and autophagy infected epithelial cells. Vitamin D levels in the blood regulate macrophages' ability to generate cathelicidin.

Additionally, vitamin D appears to influence another intrinsic antibacterial compound, called β-defensin 2. β-defensin 2 induces the release of antiviral cytokines and chemokines that recruit macrophages, neutrophils, natural killer cells, and T cells, resulting in strong host defense [22]. Vitamin D can also serve as an antibacterial by regulating cellular iron metabolism. The survival of bacteria is dependent on intracellular iron. During an infection, hepcidin is formed, which blocks transcellular iron export via ferroportin, increasing cellular iron levels. Vitamin D is a potent hepcidin suppressor, increasing ferroportin and decreasing intracellular iron levels, inhibiting bacterial growth [23].

5. Vitamin D as a regulator of adaptive immune response

The adaptive immune system responds to threats based on the characteristics of the adversary, hence the name. In comparison to the innate immune response, it takes time to mature. A unique type of cell triggers this form of immune response called an antigen-presenting cell (APC). These cells next activate B and T lymphocytes, which are in charge of subsequent antigen identification. After a priming period, T and B cells are activated in body tissues such as lymph nodes, often located far from the original antigenic substance exposure site. The proliferation of activated T and B cells and post-translational alterations of immunoglobulin synthesis utilize this time, allowing the cell to respond uniquely to an antigen. An antigen's context and the surrounding environment, including serum vitamin D levels, determine the type of T-cell activated, CD4 or CD8, or within the helper T-cell class Th1, Th2, Th17, or T-regulatory cells

(Treg). There is evidence that vitamin D suppresses inflammatory T-cells and increases regulatory T-cell activity (Tregs) [24]. Tregs are a critical component of the immune system. Its essential purpose is to reduce the immunological response after removing the threat from the body. Because the immune response persists after the threat has been eradicated, collateral damage to healthy cells and tissues may occur. Tregs are crucial in COVID-19 for preventing the cytokine storm associated with severe respiratory distress. Additionally, vitamin D is a potent suppressor of macrophage activation mediated by IFN-γ, which is likely to play a role in COVID-19 [25].

6. COVID-19 and its link to vitamin D and cytokine storm

As previously acknowledged, vitamin D deficiency has been found as a possible risk factor for COVID-19. Vitamin D acts as an immunological modulator, and its lack results in a reduced innate immune system. COVID-19's severity is determined by an individual's immune system's state. This virus severely impacts individuals with weakened immune systems, whilst those with more robust immune systems remain asymptomatic or exhibit mild-to-moderate symptoms. As a result, we could establish a link between vitamin D deficiency and the severity of COVID-19 [26].

Vitamin D's antibacterial activity was found three decades ago [27]. Besides antibacterial function, vitamin D promotes antiviral activity, which is particularly relevant in COVID-19. It possesses/promotes antiviral activity by preventing viruses from entering cells and inhibiting viral reproduction via induction of cathelicidin and defensins [28]. Additionally, vitamin D may increase antiviral activity via a critical cellular process called autophagy. Autophagy contributes to the creation of a hostile cellular environment for viruses such as coronavirus [29].

In some individuals, COVID-19 infection is accompanied by a "cytokine storm," in which several pro-inflammatory cytokines, such as IL-1, colony-stimulating factor (granulocyte CSF, macrophage CSF, granulocyte-macrophage CSF), and tumor necrosis factor-alpha (TNF-α), are released in large quantities, allowing for the recruitment of additional immune cells to the site of injury. The coronavirus causes damage by infecting both the upper and lower airways, leading to fast viral replication and extensive infiltration of immune cells, resulting in a significantly increased release of pro-inflammatory cytokines and chemokines. This results in a condition known as acute respiratory distress syndrome (ARDS) [30]. The epithelial cells of infected airways secrete cytokines that further disrupt the innate immune system. Furthermore, it stimulates the inflow of inflammatory cells such as monocytes, neutrophils, and macrophages, exposing healthy lung cells to apoptosis [31]. The lungs' microvascular and alveolar epithelial barriers are damaged by apoptosis, contributing to vascular leakage and alveolar edema. In addition, T-cell response to viral clearance is slowed, which reduces their capacity to regulate cytokine storms [32].

Vitamin D is critical for regulating the pathophysiological manifestations of the cytokine storm. Along with 1,25-$(OH)_2$D and CYP27B1, vitamin D receptors are expressed on the airway epithelia. In addition, pulmonary alveolar macrophages express vitamin D receptors as well as CYP27B1. As a result, by boosting/activating the innate immune system and increasing local 1,25-$(OH)_2$D synthesis, vitamin D can slow down the cytokine storm and chemokine production. This increases viral neutralization and clearance while moderating pro-inflammatory reactions that follow. Vitamin D also inhibits the adaptive immune system from becoming hyperactive, facilitating a quick response to viral loads [22].

6.1 Clinical evidence between COVID-19 and vitamin D

UV-B radiations are negligible at the earth's surface throughout the winter at latitudes over 40°. In a study, Rhodes et al. discovered a link between COVID-19 deaths and countries based on latitude. He discovered that countries with latitudes less than 35° North have low death rates. This could be related to the fact that persons living in regions north of 35° North do not receive enough sunlight during the winter, resulting in vitamin D insufficiency [33].

According to D'Avolio et al., 107 numbers of COVID-19 nasopharyngeal swabs were obtained from March 1 to April 14, 2020 [34]. He determined that those who tested positive for the virus suffer from severe vitamin D insufficiency. In contrast, those who tested negative have an average or insufficient vitamin D level in their blood. The median 25-OHD concentration in the 27 individuals tested positive for SARS-CoV-2 was 11.1 ng/ml, compared to 24.6 ng/ml in those who tested negative. Gennari et al. reported reduced 25-OHD levels in COVID-19 patients hospitalized in Italy [35]. Bergman et al. performed a meta-analysis of randomized controlled trials (RCTs) evaluating vitamin D to prevent respiratory tract infections (RTI). They established that preventive vitamin D treatment decreases the risk of developing RTIs [36].

Ahmed's research incorporated bioinformatics and system biology techniques to gain a deeper understanding of the SARS-CoV-2-induced cytokine storm. Vitamin D inhibits the synthesis of pro-inflammatory cytokines by inhibiting the TNF-α induced NFkB1 signaling network. It activates the IFN-α induced Jak–STAT signaling route, promoting interferon-alpha stimulating genes (ISGs) for antiviral defense [11].

According to Daneshkhah et al., Italy, Spain, and France have the highest COVID-19 age-specific case fatality rate as well as the lowest 25OHD levels (0.25 ng/l) compared to other countries. Italians and Spaniards are heavily affected by COVID-19 and have the highest rate of hypovitaminosis D in Europe (142 according to endocrinology) [37]. According to a study of 700 Italian women aged 60 to 80, 76% had 25-OHD levels less than 12 ng/ml. (154). Over 80% of postmenopausal women have hypovitaminosis D in summer, whereas only 32% have it during the winter [38, 39].

Lau et al. evaluated the levels of 25OHD in twenty COVID-19 patients admitted to an intensive care unit. Vitamin D insufficiency was detected in 11 participants and all patients under the age of 75. Seven of these individuals exhibited 25OHD concentrations of at least 10 ng/ml. According to this study, vitamin D deficiency worsened COVID-19 infection [40]. The highest rates of vitamin D deficiency were associated with the highest rates of infection and mortality, according to a study of COVID-19 severity in Europe. Further, a preliminary investigation conducted in the United States revealed a high correlation between vitamin D deficiency and mortality and other risk factors for adverse outcomes [12]. Carpegnano et al. demonstrated in a retrospective and observational investigation that vitamin D insufficiency predicts poor prognosis in patients with acute respiratory failure caused by COVID-19 [41].

In COVID-19 patients, vitamin D deficiency is associated with a high risk of hospitalization. A team of investigators undertook a recent study on elderly patients on a sample of 105 persons aged 65 and up who displayed COVID-19-like symptoms. They discovered that 35 (33.3%) of the participants tested negative for SARS-CoV-2 (vitamin D level = 52 nmol/l) when tested. Whether or not, 70 (66.7%) of individuals tested positive for SARS-CoV-2 (vitamin D level = 27 nmol/l). Among the 70 (66.7%) people, 39 (55.7%) had vitamin D levels below 30, whereas the remaining 31 (44.3%) had Vitamin D levels above 30, and their D-dimer value was lower (D-dimer is considered a risk factor for blood clots in COVID-19) [42].

The production of IL-6 by monocytes, dendritic cells, and macrophages during COVID-19 infection resulted in producing a large number of pro-inflammatory cytokines and C-reactive protein (CRP). CRP is a nonspecific inflammatory marker that becomes more specific to the bioactivity of IL-6. It can generate significant inflammation, which is critical in the development of cytokine storms in individuals with severe COVID-19 [43]. Guan et al. reported that approximately 44.5% of severe COVID-19 patients with high CRP levels had a higher risk than patients with mild COVID-19 [44]. Vitamin D deficiency promotes the release of cytokines such as TNF-α and IL-1, leading to increased inflammation and elevated CRP [45]. This could explain the concurrent reduction in CRP and inflammatory cytokines ($CD4^+$ and IFN) in hemodialysis patients following calcitriol treatment, as well as the elevation of both CRP and cytokines in severe COVID-19 patients. Experiments have shown that vitamin D can inhibit the production of inflammatory cytokines such as IL-6 and L-17 while increasing anti-inflammatory cytokines such as IL-10 [46]. This could lead to a reduction in CRP and inflammation.

Studies show that vitamin D deficiency can exacerbate RTI in vulnerable populations (elderly and those with underlying health conditions). To maintain an adequate vitamin D status and prevent mortality, supplementation is recommended to restore vitamin D levels depleted due to insufficient sunlight exposure. The clinical data on vitamin D status on COVID-19 disease is still in the initial stage. As of now, many of the papers in this field have been published without extensive peer review, retrospectively, and with only associative data [47].

7. Supplemental vitamin D acts as an anti-inflammatory agent

By regulating the production of inflammatory cytokines and inhibiting the growth of pro-inflammatory cells, vitamin D aids in modulating the immune system [48]. According to various studies assessing the association between low vitamin D levels and acute infections, vitamin D supplementation improves clinical responses to acute infections. In addition to its role in natural cellular immunity, vitamin D also plays a role in physical barriers and adaptive immunity [49]. Several studies have shown that vitamin D influences the adaptive immune response in inflammatory and autoimmune diseases [50, 51]. In addition, it may play a role in the etiology of chronic inflammatory disorders, such as asthma, atherosclerosis-associated cardiovascular disease, inflammatory bowel disease, nonalcoholic fatty liver disease, and chronic renal disease [52]. During the autumn and winter months in the UK, the National Institute for Health and Care Excellence has recommended taking vitamin D supplements to prevent bone, muscle, and immunological problems [53].

Vitamin D intake reduces the likelihood of contracting influenza, according to a study [54]. More research, however, is needed to confirm these findings. Vitamin D supplementation aids in HIV infection by blocking viral entry, regulating $CD4^+$ cell surface antigen expression, dampening viral p24 production, and reducing monocyte proliferation [55]. A deficit of vitamin D is associated with acute respiratory distress syndrome, chronic illness, and increased mortality rates in the elderly [56]. An experimental report indicates that high-dose vitamin D supplementation may be beneficial, especially for the elderly, obese, those with dark skin, those living at higher latitudes, and those living at 35° north, where sunlight intake is insufficient to maintain adequate vitamin D levels during winter. A vitamin D supplement may be beneficial to people at risk of chronic diseases such as respiratory tract infections, cardiovascular disease, cancer, diabetes, and hypertension. In the bloodstream, levels of vitamin D above 50 ng/ml (125 nmol/l) have been linked to a lower risk of many viral infections, including COVID-19 [57, 58].

8. Conclusion

Based on recent clinical investigations, we conclude that vitamin D may have a role in lowering the complications associated with COVID-19-induced uncontrolled inflammation and cytokine storm. Additionally, researchers found a link between patients with COVID-19 and insufficiency of serum vitamin D. The widespread effects of vitamin D on various organ systems have sparked discussion regarding a possible interaction between it and the processes by which the SARS-CoV-2 virus infects humans. In COVID-19 patients, adequate vitamin D levels in serum were associated with a significantly lower frequency of disease severity, despite limited clinical data. It is expected that there will be more studies published regarding its role in upper respiratory infections such as COVID-19 in the near future. A sufficient amount of patient-level data is required to demonstrate the protective effect of vitamin D.

Author contributions

AB, BD & HS: Conceptualization, Super vision, Writing- review & editing; AA, SK, SK, BB, SS & RD: Data collection, Writing.

Funding

There was no funding for this project from any government, commercial, or non-profit organization.

Conflict of interest

We have no conflict of interest.

Declaration of interest

The authors state that they have no conflicts of interest that could impede the impartiality of this chapter.

List of abbreviations

ACE II	Angiotensin-Converting Enzyme II
APC	Antigen-Presenting Cell
ARDS	Acute Respiratory Distress Syndrome
CD-4 & 10	Cluster of Differentiation-4 & 10
COVID-19	Coronavirus disease-19
CRP	C-reactive protein

DOI: http://dx.doi.org/10.5772/intechopen.106980

DHCR7	7-Dehydrocholesterol Reductase
HIV	Human Immunodeficiency Virus
IFN-γ	Interferon gamma
IL	Interleukin
MERS	Middle East Respiratory Syndrome
RBD	Receptor-Binding Domains
RCTs	Randomized Controlled Trials
RNA	Ribonucleic acid
RTI	Respiratory Tract Infections
SARS	Severe Acute Respiratory Syndrome
SARS-CoV-2	Severe Acute Respiratory Syndrome Coronavirus-2
TMPRSS2	Transmembrane Protease Serine-2
TNF-α	Tumor Necrosis Factor-alpha
UK	United Kingdom
UV-B	Ultraviolet B
VDR	Vitamin D Receptor

Author details

Alakesh Bharali[1], Bhargab Deka[2*], Himangshu Sarma[3,4*], Ashique Ahmed[5], Bedanta Bhattacharjee[6], Santa Sarma[1], Suman Kumar[7], Susankar Kushari[7] and Rajlakshmi Devi[8]

1 Department of Pharmaceutics, Girijananda Chowdhury Institute of Pharmaceutical Science, Guwahati, Assam, India

2 Department of Pharmacology, Girijananda Chowdhury Institute of Pharmaceutical Science, Guwahati, Assam, India

3 Sophisticated Analytical Instrument Facility (SAIF), Girijananda Chowdhury Institute of Pharmaceutical Science, Guwahati, Assam, India

4 Faculty of Science and Engineering, Department of Pharmaceutical Sciences, Dibrugarh University, Dibrugarh, Assam, India

5 Department of Pharmacology, NEPDES College of Pharmaceutical Sciences, Guwahati, Assam, India

6 Faculty of Pharmaceutics, Girijananda Chowdhury Institute of Pharmaceutical Science, Tezpur, India

7 Department of Pharmaceutical Chemistry, Girijananda Chowdhury Institute of Pharmaceutical Science, Guwahati, Assam, India

8 Life Sciences Division, Institute of Advanced Study in Science and Technology, Guwahati, Assam, India

*Address all correspondence to: bhargavdeka98@gmail.com; himangshu.amc@gmail.com

References

[1] World Health Organization. WHO Coronavirus (COVID-19) Dashboard. 2022. Available from: https://covid19.who.int/ [Accessed May 13, 2022]

[2] Huang C, Wang Y, Li X, Ren L, Zhao J, Hu Y, et al. Clinical features of patients infected with 2019 novel coronavirus in Wuhan, China. Lancet. 2020;**395**(10223):497-506

[3] Zhu N, Zhang D, Wang W, Li X, Yang B, Song J, et al. A novel coronavirus from patients with pneumonia in China, 2019. New England Journal of Medicine. 2020;**382**(8):727-733

[4] Gao J, Mei H, Sun J, Li H, Huang Y, Tang Y, et al. Neuropilin-1-mediated SARS-CoV-2 infection in bone marrow-derived macrophages inhibits osteoclast differentiation. Advanced Biology. 2022;**6**:1-14

[5] Morris A. Effects of pancreatic SARS-CoV-2 infection identified. Nature Reviews Endocrinology. 2021;**17**(4):192-192

[6] Cheung KS, Hung IFN, Chan PPY, Lung KC, Tso E, Liu R, et al. Gastrointestinal manifestations of SARS-CoV-2 infection and virus load in fecal samples from a Hong Kong cohort: Systematic review and Meta-analysis. Gastroenterology. 2020;**159**(1):81-95

[7] Mao L, Jin H, Wang M, Hu Y, Chen S, He Q, et al. Neurologic manifestations of hospitalized patients with coronavirus disease 2019 in Wuhan, China. JAMA Neurology. 2020;**77**(6):683-690

[8] Deka B, Dash B, Newar A, Ahmed A, Bharali A, Bordoloi B, et al. Coronavirus disease 2019: A comprehensive review on recent clinical updates and management tools for pharmacists. Pharmaceutical and Biosciences Journal. 2021;**9**(3):01-21

[9] Ragab D, Salah Eldin H, Taeimah M, Khattab R, Salem R. The COVID-19 cytokine storm; what we know so far. Frontiers in Immunology. 2020;**11**:1446

[10] Razdan K, Singh K, Singh D. Vitamin D levels and COVID-19 susceptibility: Is there any correlation? Medicine in Drug Discovery. 2020;**7**:100051

[11] Ahmed F. A network-based analysis reveals the mechanism underlying vitamin D in suppressing cytokine storm and virus in SARS-CoV-2 infection. Frontiers in Immunology. 2020;**11**:3084

[12] Daneshkhah A, Agrawal V, Eshein A, Subramanian H, Roy HK, Backman V. The possible role of vitamin D in suppressing cytokine storm and associated mortality in COVID-19 patients. medRxiv. 2020:2020.04.08.20058578. DOI: 10.1101/2020.04.08.20058578

[13] Cleary SJ, Pitchford SC, Amison RT, Carrington R, Robaina Cabrera CL, Magnen M, et al. Animal models of mechanisms of SARS-CoV-2 infection and COVID-19 pathology. British Journal of Pharmacology. 2020;**177**(21):4851-4865

[14] Ziegler CGK, Allon SJ, Nyquist SK, Mbano IM, Miao VN, Tzouanas CN, et al. SARS-CoV-2 receptor ACE2 is an interferon-stimulated gene in human airway epithelial cells and is detected in specific cell subsets across tissues. Cell. 2020;**181**(5):1016-1035.e19

[15] Song Z, Bao L, Yu P, Qi F, Gong S, Wang J, et al. SARS-CoV-2 causes a systemically multiple organs damages

and dissemination in hamsters. Frontiers in Microbiology. 2021;**11**:3451

[16] Bishop LE, Ismailova A, Dimeloe S, Hewison M, White JH. Vitamin D and immune regulation: Antibacterial, antiviral, anti-inflammatory. JBMR Plus. 2020;**5**(1):e10405

[17] Fabbri A, Infante M, Ricordi C. Vitamin D status: A key modulator of innate immunity and natural defense from acute viral respiratory infections. European Review for Medical and Pharmacological Sciences. 2020;**24**(7):4048-4052

[18] Laird E, Rhodes J, Kenny RA. Vitamin D and inflammation: Potential implications for severity of Covid-19. Irish Medical Journal. 2020;**113**(5):81

[19] Martens PJ, Gysemans C, Verstuyf A, Mathieu C. Vitamin D's effect on immune function. Nutrients. 2020;**12**(5):1248

[20] Baeke F, Takiishi T, Korf H, Gysemans C, Mathieu C. Vitamin D: Modulator of the immune system. Current Opinion in Pharmacology. 2010;**10**(4):482-496

[21] Koivisto O, Hanel A, Carlberg C. Key vitamin D target genes with functions in the immune system. Nutrients. 2020;**12**(4):1140

[22] Bilezikian JP, Bikle D, Hewison M, Lazaretti-Castro M, Formenti AM, Gupta A, et al. Mechanisms in endocrinology: Vitamin D and COVID-19. European Journal of Endocrinology. 2020;**183**(5):R133-R147

[23] Ebadi M, Montano-Loza AJ. Perspective: Improving vitamin D status in the management of COVID-19. European Journal of Clinical Nutrition. 2020;**74**(6):856-859

[24] Singh P, Kumar M, Al KS. Vitamin D deficiency in the Gulf cooperation council: Exploring the triad of genetic predisposition, the gut microbiome and the immune system. Frontiers in Immunology. 2019;**10**(MAY):1042

[25] Bikle DD, Vitamin D. Regulation of immune function. In: Litwack G, editor. Vitamins and Hormones. Cambridge, Massachusetts: Academic Press; 2011. pp. 1-21

[26] Hewison M. Vitamin D and the intracrinology of innate immunity. Molecular and Cellular Endocrinology. 2010;**321**(2):111

[27] Rook GAW, Steele J, Frahert L, Barkert S, Karmalit R, O'riordant J, et al. Vitamin D3, gamma interferon, and control of proliferation of mycobacterium tuberculosis by human monocytes. Immunology. 1986;**57**(1):163

[28] Barlow PG, Svoboda P, Mackellar A, Nash AA, York IA, Pohl J, et al. Antiviral activity and increased host Defense against influenza infection elicited by the human cathelicidin LL-37. PLoS One. 2011;**6**(10):e25333

[29] Mao J, Lin E, He L, Yu J, Tan P, Zhou Y. Autophagy and viral infection. Autophagy Regulation of Innate Immunity. 2019;**1209**:78

[30] Channappanavar R, Perlman S. Pathogenic human coronavirus infections: Causes and consequences of cytokine storm and immunopathology. Seminars in Immunopathology. 2017;**39**(5):529-539

[31] Channappanavar R, Fehr AR, Vijay R, Mack M, Zhao J, Meyerholz DK, et al. Dysregulated type I interferon and inflammatory monocyte-macrophage responses cause lethal pneumonia in SARS-CoV-infected mice. Cell Host & Microbe. 2016;**19**(2):181-193

[32] Zhao J, Zhao J, Perlman S. T cell responses are required for protection from clinical disease and for virus clearance in severe acute respiratory syndrome coronavirus-infected mice. Journal of Virology. 2010;**84**(18):9318-9325

[33] Rhodes JM, Subramanian S, Laird E, Kenny RA. Low population mortality from COVID-19 in countries south of latitude 35 degrees north supports vitamin D as a factor determining severity. Alimentary Pharmacology & Therapeutics. 2020;**51**(12):1434-1437

[34] D'avolio A, Avataneo V, Manca A, Cusato J, De Nicolò A, Lucchini R, et al. 25-hydroxyvitamin D concentrations are lower in patients with positive PCR for SARS-CoV-2. Nutrients. 2020;**12**(5):1359

[35] Gennari L, Campi I, Merlotti D, Mingiano C, Frosali A, Giovanelli L, et al. Vitamin D deficiency is independently associated with COVID-19 severity and mortality. Journal of Bone and Mineral Research. 2020;**35**:7-8

[36] Bergman P, Lindh ÅU, Björkhem-Bergman L, Lindh JD. Vitamin D and respiratory tract infections: A systematic review and Meta-analysis of randomized controlled trials. PLoS One. 2013;**8**(6):e65835

[37] Daneshkhah A, Eshein A, Subramanian H, Roy HK, Backman V. The role of vitamin D in suppressing cytokine storm in COVID-19 patients and associated mortality. medRxiv. 2020:2020.04.08.20058578. DOI: 10.1101/2020.04.08.20058578

[38] Hossein-Nezhad A, Holick MF. Vitamin D for health: A global perspective. Mayo Clinic Proceedings. 2013;**88**(7):755

[39] Romagnoli E, Caravella P, Scarnecchia L, Martinez P, Minisola S. Hypovitaminosis D in an Italian population of healthy subjects and hospitalized patients. British Journal of Nutrition. 1999;**81**(2):133-137

[40] Lau FH, Majumder R, Torabi R, Saeg F, Frank Lau RH, Hoffman R, et al. Vitamin D insufficiency is prevalent in severe COVID-19. medRxiv. 2020: 2020.04.24.20075838. DOI: 10.1101/2020.04.24.20075838

[41] Carpagnano GE, Di Lecce V, Quaranta VN, Zito A, Buonamico E, Capozza E, et al. Vitamin D deficiency as a predictor of poor prognosis in patients with acute respiratory failure due to COVID-19. Journal of Endocrinological Investigation. 2021;**44**(4):765-771

[42] Baktash V, Hosack T, Patel N, Shah S, Kandiah P, Van Den Abbeele K, et al. Vitamin D status and outcomes for hospitalised older patients with COVID-19. Postgraduate Medical Journal. 2021;**97**(1149):442-447

[43] Moore JB, June CH. Cytokine release syndrome in severe COVID-19. Science. 2020;**368**(6490):473-474

[44] Guan W, Ni Z, Hu Y, Liang W, Ou C, He J, et al. Clinical characteristics of coronavirus disease 2019 in China. New England Journal of Medicine. 2020;**382**(18):1708-1720

[45] Daneshkhah A, Agrawal V, Eshein A, Subramanian H, Roy HK, Backman V. Evidence for possible association of vitamin D status with cytokine storm and unregulated inflammation in COVID-19 patients. Aging Clinical and Experimental Research. 2020;**32**(10):2141-2158

[46] McGregor R, Chauss D, Freiwald T, Yan B, Wang L, Nova-Lamperti E, et al. An autocrine vitamin D-driven

Th1 shutdown program can be exploited for COVID-19. bioRxiv. 2020:2020.07.18.210161. DOI: 10.1101/2020.07.18.210161

[47] Chakhtoura M, Napoli N, El Hajj FG. Commentary: Myths and facts on vitamin D amidst the COVID-19 pandemic. Metabolism. 2020;**109**:154276

[48] White JH. Vitamin D signaling, infectious diseases, and regulation of innate immunity. Infection and Immunity. 2008;**76**(9):3837-3843

[49] Rondanelli M, Miccono A, Lamburghini S, Avanzato I, Riva A, Allegrini P, et al. Self-care for common colds: The pivotal role of vitamin D, vitamin C, zinc, and Echinacea in three main immune interactive clusters (physical barriers, innate and adaptive immunity) involved during an episode of common colds—Practical advice on dosages and on the time to take these nutrients/botanicals in order to prevent or treat common colds. Evidence-Based Complementary and Alternative Medicine. 2018;**2018**:5813095

[50] Olliver M, Spelmink L, Hiew J, Meyer-Hoffert U, Henriques-Normark B, Bergman P. Immunomodulatory effects of vitamin D on innate and adaptive immune responses to Streptococcus pneumoniae. Journal of Infectious Diseases. 2013;**208**(9):1474-1481

[51] Tiosano D, Wildbaum G, Gepstein V, Verbitsky O, Weisman Y, Karin N, et al. The role of vitamin D receptor in innate and adaptive immunity: A study in hereditary vitamin D-resistant rickets patients. Journal of Clinical Endocrinology and Metabolism. 2013;**98**(4):1685-1693

[52] Yin K, Agrawal DK. Vitamin D and inflammatory diseases. Journal of Inflammation Research. 2014;**7**(1):69-87

[53] Ulivieri FM, Banfi G, Camozzi V, Colao A, Formenti AM, Frara S, et al. Vitamin D in the Covid-19 era: A review with recommendations from a GIOSEG expert panel. Endocrine. 2021;**72**(3):597-603

[54] Gruber-Bzura BM. Vitamin D and influenza—Prevention or therapy? International Journal of Molecular Sciences. 2018;**19**(8):2419

[55] Mansueto P, Seidita A, Vitale G, Gangemi S, Iaria C, Cascio A. Vitamin D deficiency in HIV infection: Not only a bone disorder. BioMed Research International. 2015;**2015**:735615

[56] Grant WB, Lahore H, McDonnell SL, Baggerly CA, French CB, Aliano JL, et al. Evidence that vitamin D supplementation could reduce risk of influenza and covid-19 infections and deaths. Nutrients. 2020;**12**(4):988

[57] Misra DP, Agarwal V, Gasparyan AY, Zimba O. Rheumatologists' perspective on coronavirus disease 19 (COVID-19) and potential therapeutic targets. Clinical Rheumatology. 2020;**39**(7):2055-2062

[58] Murdaca G, Pioggia G, Negrini S. Vitamin D and Covid-19: An update on evidence and potential therapeutic implications. Clinical and Molecular Allergy. 2020;**18**(1):23

Chapter 6

Role of Vitamin D in Patients with Schizophrenia Suffering from COVID-19

Fatemeh Gholami, Saman Farshid, Parmida Soleimani and Rohollah Valizadeh

Abstract

People with schizophrenia are at high risk for vitamin D deficiency. There is more likely as association between vitamin D and COVID-19 development and even severe outcomes following SARS-CoV-2 infection. It should be noted that other factors except schizophrenia are also related to the severity of the COVID-19 such as heart conditions, respiratory disorders, overweight, and hypertension in which are prevalent in patients with schizophrenia linked with vitamin D deficiency. This book aimed to determine the relationship between the level of vitamin D and COVID-19 severity in patients with schizophrenia.

Keywords: schizophrenia, vitamin D, 25(OH) D, respiratory infection, COVID-19, coronavirus disease

1. Introduction

In December 2019, severe acute respiratory syndrome coronavirus 2 (SARS-CoV-2) emerged in Wuhan, Hubei province, China. The World Health Organization (WHO) declared coronavirus disease 2019 (COVID-19) a pandemic on 11 March 2020. While it is estimated that 80% of those infected with COVID-19 are asymptomatic or have a self-limiting disease, the case fatality rate for those hospitalized with COVID-19 was 2.3%, increasing to 10.5% in those with cardiovascular disease, 7.3% in those with diabetes mellitus, 6% in those with hypertension and 5.6% for cancer [1]. COVID-19 was distributed worldwide and showed various symptoms, including lung involvement, liver and kidney damage, and conjunctivitis. COIVD-19 is considered a disease with multi-organ failure ability [2–7]. After declaring the Pandemic by WHO, researches were done further to find remedy and vaccines [8–10]. Vitamin D is one of the subjects that had controversial effects on the treatment or recovery process of patients with COVID-19. Now we elaborate on the details of vitamin D [11].

Vitamin D, a steroid hormone, plays the main role in the immune system [12, 13]. Vitamin D influences many reactions against the normal immune response to pathogens. Vitamin D can facilitate the recovery of COVID-19 because both cytokine storms and inflammation are related to severe outcomes in patients with COVID-19 who have high prevalence of pneumonia and lung failure, especially in older patients with lots of comorbidities results in high mortality rate. More than 70% of all schizophrenia (SCZ) patients also have one or more clinical conditions, including diabetes type II [14–16], chronic pulmonary disease [17], heart diseases, obesity, and hypertension, so the life span in these people decreases [18–20] and may be vulnerable to infection with SARA-CoV-2 [21].

People who have psychotic disorders are at high risk for vitamin D deficiency [22]. There is a close relationship between COVID-19 and SCZ. The results of a study showed that SCZ is associated with high mortality following COVID-19 development [23].

Low accessibility to suitable medical care aggravates this scenario [24]. Patients with SCZ and their home care providers may have problem seeking health services. Additionally, even if they want to ask for medical assistance, due to the stigma surrounding SCZ, there is more likely to not take proper assessment or treatment [25].

Patients with schizophrenia are prone to be infected with worse outcomes, especially if they are suffering from several comorbidities. They are vulnerable to worsening psychiatric symptoms and relapse due to fear of the disease, stress, and the boredom associated with compulsory isolation. Thus, health and care providers need more attention and support to prevent COVID-19 among in this group and should detect both psychiatric and respiratory problems as soon as possible.

In this chapter, we tried to describe the reason for high COVID-19 morbidity and mortality among individuals with SCZ through a literature review.

2. Development of vitamin D deficiency in patients with schizophrenia

Schizophrenia may be developed by environmental and genetic factors [26]. According to epidemiological research, schizophrenia is seen more in people: 1) born in the winter and spring seasons [27], 2) living in the urban area in childhood [28–31], and 3) living at high latitudes [32]. On the other hand, we know that dark skin needs much sunlight exposure to produce enough vitamin D, so children with dark skin who migrate to cold climates have more chance of developing SCZ due to low levels of vitamin D during gestation [33]. A Danish case–control study indicated that vitamin D deficiency in neonates is associated with an increased risk for SCZ in later life. People with 25OHD less than 20.4 nmol/L [34] had a 44% increased risk of SCZ compared to people over 40.1–53.5 nmol/L [22].

However, randomized clinical trials to investigate the effects of maternal vitamin D supplements on the development of SCZ in their children may never happen due to two reasons. First, there is no strong evidence suggesting high dose of 25OHD for

Study	Hypothesis
Finnish birth cohort	Lack of vitamin D supplements in the first year of life is associated with SCZ development in men [36].
Danish case–control study	Vitamin D deficiency in neonates is associated with SCZ development in later life [22].
KiGGS[1] study	The offspring of migrants with dark skin develop more SCZ [33].

[1]*The German National Health Interview and Examination Survey for Children and Adolescents.*

Table 1.
Summary of hypothesis related to vitamin D and SCZ.

health targets. Second, it is unethical to screen pregnant women for vitamin D deficiency and allocate this group to take vitamin D supplements or the placebo. Finally, it is hard to follow up on large mother-offspring samples for 2–3 decades to determine the risk of SCZ development [35]. A summary of the hypothesis related to vitamin D and SCZ is presented in **Table 1**.

3. Prevalence of vitamin D deficiency in schizophrenia

Vitamin D deficiency and insufficiency are in different people worldwide, but its burden, such that the prevalence of serum 25OHD < 25/30 nmol/L ranges from 5 to 18%, depends on the Food and Agriculture Organization (FAO) world region, varies from 24 to 49% in the case of serum 25OHD < 50 nmol/L [37]. It is not clear which dose of vitamin D levels are sufficient, insufficient, and deficient, and we showed typical thresholds in **Table 2**. The main circulating form of vitamin D is 25OHD which is usually taken as a proxy of vitamin D status in blood [39, 40]. Two prominent studies report an association between neonatal vitamin D deficiency and an increased risk of SCZ [26, 41]. A meta-analysis study including 31 studies showed that there were statistically significant differences in the mean 25OHD between SCZ and the control group in which; the control group in the case–control and cohort studies consisted of healthy subjects with no history of psychiatric disorders, while in the cross-sectional studies, psychiatric patients but non-schizophrenic were considered to be the control group. Consequently, it can be concluded that, compared with healthy people or other psychiatric patients, peripheral blood mean 25OHD is low in patients with SCZ [26].

Generally, patients suffering from SCZ have poor general health, poor nutrition, low activity, and more comorbidity. So it is important to be cautious in any causal interpretation for patients with SCZ [42]. Some studies have inconsistent results from existing vitamin D supplementation trials in patients with SCZ [43, 44]. We can conclude that vitamin D deficiency can increase the risk of SCZ development, and it is strongly recommended to do ongoing research.

4. Vitamin D deficiency and respiratory infection risk with SARS-CoV-2 virus

There are two critical questions emanating from the title of this part. The first is whether there is an association between susceptibility to develop COVID-19 and vitamin D deficiency or not. The second is whether vitamin D use in deficient people can prevent or improve infection with SARS-CoV-2 or change the course of its disease.

Serum 25(OH)D concentration, mean: nmol/L (ng/mL)	Prevalence, %	Classification
<25 (<10)	6.7	Deficiency
<50 (<20)	37	Insufficiency
>75 (>30)	11	Sufficiency

25(OH)D, 25-hydroxyvitamin D [38].

Table 2.
Vitamin D level and its prevalence worldwide.

Improvement of the immune system by correct nutrition is a considerable factor. Vitamin D as a nutrient plays a significant role in the immune system. However, there is little evidence about the role of vitamin D in preventing COVID-19 and its consequences [45].

COVID-19 pandemic has raised challenges in terms of the benefits of vitamin D used to prevent and even treat it. Sufficient blood vitamin D can help in a satisfactory cellular response and in protecting against the severity of infections caused by microorganisms such as SARS-CoV-2 [45]. Vitamin D deficiency is related to severe outcomes following COVID-19 [46]. In a systemic review and meta-analysis consisting of 14 studies, the effect of vitamin D supplementation in lowering the risk of non-COVID-19 respiratory tract infections in patients with lower vitamin D levels was found [47]. Moreover, a systematic review consisting of 7 meta-analyses on 30 clinical trials showed the same results [48]. Higher COVID-19 mortality rates in Europe have been identified in patients suffering from vitamin D deficiency [45].

There is still limited evidence in favor of the effect of vitamin D in people with COVID-19 in the treatment process [49]. A meta-analysis consisting of eight observational studies revealed that people with vitamin D < 50 nmol/l (i.e., <20 ng/ml) have 64% more risk of community-acquired pneumonia [50].

The results of the meta-analysis showed that vitamin D deficiency could result in the severe form of COVID-19, especially in the elder people [49], which is explained by both lower exposure to sunlight and lower 7-dehydrocholesterol values in the skin compromising the cutaneous synthesis of 25OHD in the elderly [51]. Moreover, aging is accompanied by lots of chronic diseases [52].

We recommend developing cohort studies, especially clinical trials on different age groups in various climatic conditions, to evaluate the causality between vitamin D and COVID-19.

4.1 Mechanism: vitamin D regulating inflammation

The association of vitamin D and C-reactive protein (CRP) level, an anti-inflammatory factor, is proposed. Vitamin D use is associated with a reduction in CRP level [53], while in patients with SCZ, an inverse relationship was found between CRP levels and vitamin D [54]. Low serum level of vitamin D seems to be associated with the inappropriate function of the immunomodulatory. Also, insufficient vitamin D levels result in less efficient antigen presentation and macrophage function. Low vitamin D may potentially contribute to a delayed response to the body's initial contact with the SARS-CoV-2 virus.

5. Schizophrenia and COVID-19

A large cohort study was carried out on 1092 patients with/without SCZ hospitalized due to COVID-19. Only 15 patients had SCZ. The overall in-hospital mortality rate was 9%. Patients with CSZ had more mortality compared to non-SCZ patients (26.7% vs. 8.7%) (Adjusted odds ratio: 4.36). We know that an adjusted odds ratio of more than one is considered a risk factor, while here, it is 4.36 [23]. Vitamin D deficiency is associated with higher risk of respiratory infection. There are more respiratory infections and deaths in patients with SCZ where vitamin D deficiency is prevalent. This potentially offers a modifiable risk factor to reduce the risk for and the severity of respiratory infection in people with SCZ [21].

5.1 Prognostic factors in developing COVID-19

5.1.1 Age and gender

In spite of the fact that females experience more morbidity (not mortality) than males [54], the relative risk of mortality following COVID-19 was higher for males than females in all regions and almost all age groups. The overall relative risk ranged from 1.11 in Portugal to 1.54 in France, showing the risk factor role of gender in COVID-19 consequences. In most regions, sex differences increase until 60–69 years [55]. Clinicians obviously noted that COVID-19 mortality rises with aging, unlike other respiratory diseases [56]. People ≥65 years have a strikingly higher mortality rate following COVID-19 compared to younger adults, and males have a higher mortality rate following COVID-19 compared to females. Over the 42-day period, there were 178,568 deaths following COVID-19 deaths. Mortality was influenced by age and sex in patients with COVID-19. Compared with individuals≤54 years, the incident rate ratio [57] was 8.1, indicating the high mortality rate following COVID-19, and also IRR was 62.1 for patients≥65 years or older. Totally mortality rate due to COVID-19 was 77% higher in males compared to females (IRR = 1.77) [58].

In addition, age may also have interaction with SCZ in respect of the mortality rate following SARS-CoV-2 infection. A retrospective case–control study showed that patients with SCZ >65 years had higher risk compared to the patients with SCZ aged 18–65 years (Adjusted odds ratio = 1.74) [59].

5.1.2 Ethnicity

In the general population, mortality following infection with SARS-CoV-2 among people from ethnic minorities is four times higher than in the white European population [60]. In an observational study, as compared to white patients, African-American people suffering from SCZ had higher prevalence of SARS-CoV-2 infection (adjusted odds ratio: 2.33) and higher mortality rate (6.2% vs. 3.7%), and men had higher mortality rate than women (6.6% vs. 3.4%) [61].

5.1.3 Comorbidities

The risk factors for severe SARS-CoV-2 infection, such as cardiovascular disease (CVD), chronic respiratory diseases such as chronic obstructive pulmonary disease (COPD) and diabetes mellitus (DM) [62–64], are frequent in patients with SCZ compared to the general people. More than 70% of all patients with SCZ have one or more comorbidity, including diabetes type II [15, 16], chronic pulmonary disease [17], hypertension, heart diseases, and obesity, so overall survival in these patients decreases [18–20] and is the vulnerable group to COVID-19 with high mortality [21]. Patients with SCZ and/or with other mental problems such as bipolar disorders had high risk of mortality following COVID-19. This can be justified by their immunological profile. Variation in the human leukocyte antigen complex is one of the most consistently replicated findings in genome-wide association studies in patients with SCZ and bipolar disorders [65]. In conclusion, the highest risk of mortality in individuals with SCZ and/or bipolar disorders is not far off [66].

5.2 Vitamin D deficiency in schizophrenia implications for COVID-19 infection

The global age-standardized prevalence of SCZ is 0.28% [67]. Among COVID-19 risk factors identified in patients with SCZ, the presence of comorbidities, stigma experience, poor insight into somatic symptoms, cognitive impairment, and delusions have been identified as factors that can lead to misperception of the risk related to the virus. Moreover, patients with SCZ are often heavy smokers with lower vitamin D levels, and it is unknown how it can affect their chance of COVID-19 survival. A case–control study on patients with COVID-19 in southern France showed that the mortality rate was 9.0%. The patients with SCZ had increased mortality compared to the non-SCZ patients (26.7% vs. 8.7%, respectively). In contrast, the patients with SCZ were admitted to the ICU less than those without SCZ. SCZ is associated with further COVID-19 mortality, confirming the existence of health disparities described in other somatic diseases [23].

Lack of vitamin D causes deterioration in the health of our body and thus increases the risk of mental disorders. Research is ongoing, but studies have shown that sunlight provides a significant protective effect for respiratory problems and inflammation disorders. In the context of the coronavirus pandemic, research has been conducted on the relationship between vitamin D, CZ, and increased rates of acute respiratory infection.

There is more respiratory infection and mortality in patients with SCZ whose vitamin D deficiency is prevalent [68]. A case series study including 14 elderly COVID-19 positive inpatients presenting with dementia or SCZ and other medical conditions was done. All patients received 800 IU daily vitamin D prior to the infection. Most of the patients were asymptomatic or with very few symptoms. There was no need for an intensive care unit, or deaths were not reported. But cognitive functioning of the patients was unchanged. It can be concluded pre-existing vitamin D use may reinforce the immune system and lower the severity of COVID-19 in elderly patients with psychiatric disorders [69].

6. Conclusions

It seems vitamin D deficiency is associated with an increased risk of acute respiratory infection and mortality after the development of COVID-19. There are further respiratory tract infections and mortality in patients with schizophrenia because vitamin D deficiency is prevalent in these patients. Patients with schizophrenia are prone to be infected with worse outcomes, especially if they suffer from several comorbidities. They are vulnerable to worsening psychiatric symptoms and relapse due to fear of the disease, stress, and the boredom associated with compulsory isolation. Thus, health and care providers need more attention and support to prevent COVID-19 in this group and should detect psychiatric and respiratory problems as soon as possible.

Author details

Fatemeh Gholami[1,2], Saman Farshid[3], Parmida Soleimani[4] and Rohollah Valizadeh[5*]

1 Department of Epidemiology, School of Public Health, Iran University of Medical Sciences, Tehran, Iran

2 Social Determinants of Health Research Center, Research Institute for Health Development, Kurdistan University of Medical Sciences, Sanandaj, Iran

3 Nephrology and Kidney Transplant Research Center, Clinical Research Institute, Urmia University of Medical Sciences, Urmia, Iran

4 Department of Psychology, Roudehen Branch, Islamic Azad University, Tehran, Iran

5 Urmia University of Medical Sciences, Urmia, Iran

*Address all correspondence to: rohvali4@gmail.com

References

[1] Wu Z, McGoogan JM. Characteristics of and important lessons from the coronavirus disease 2019 (COVID-19) outbreak in China: Summary of a report of 72 314 cases from the Chinese Center for Disease Control and Prevention. Journal of the American Medical Association. 2020;**323**:1239-1242. DOI: 10.1001/jama.2020.2648

[2] Dadashzadeh N, Farshid S, Valizadeh R, Nanbakhsh M, Mohammad RM. Acute respiratory distress syndrome in COVID-19. Immunopathologia Persa. 2020;**6**:e16. DOI: 10.34172/ipp.2020.16

[3] Lotfi B, Farshid S, Dadashzadeh N, Valizadeh R, Rahimi MM. Is coronavirus disease 2019 (COVID-19) associated with renal involvement? A review of century infection. Jundishapur Journal of Microbiology. 2020;**13**:e102899. DOI: 10.5812/jjm.102899

[4] Daneshfar M, Dadashzadeh N, Ahmadpour M, Ragati Haghi H, Rahmani V, Frouzesh M, et al. Lessons of mortality following COVID-19 epidemic in the United States especially in the geriatrics. Journal of Nephropharmacology. 2021;**10**:e06. DOI: 10.34172/npj.2021.06

[5] Tabatabaii SA, Soltani P, Khanbabaee G, Sharma D, Valizadeh R, Farahbakhsh N, et al. SARS Coronavirus 2, severe acute respiratory syndrome, and middle east respiratory syndrome in children: A review on epidemiology, clinical presentation, and diagnosis. Archives of Pediatric Infectious Diseases. 2020;**8**:e104860. DOI: 10.5812/apid.104860

[6] Besharat SAN, Dadashzadeh N, Talaie R, Mousavi SS, Barzegar A, Tavana S, et al. Clinical and demographic characteristics of patients with COVID-19 who died in Modarres Hospital. Open Access Macedonian Journal of Medical Sciences. 2020;**8**:144-149. DOI: 10.3889/oamjms.2020.5013

[7] Mirshamsi M, Ghiasi N, Heidari S, Hosseinpour P, Hassanlouei B, Hashemipour SMA, et al. Conjunctivitis and other ocular manifestation following COVID-19; updated information about transmission of COVID-19 by eye. Immunopathologia Persa. 2021;**7**:e28. DOI: 10.34172/ipp.2021.28

[8] Rahimi MM, Jahantabi E, Lotfi B, Forouzesh M, Valizadeh R, Farshid S. Renal and liver injury following the treatment of COVID-19 by remdesivir. Journal of Nephropathology. 2021;**10**:e10. DOI: 10.34172/jnp.2021.10

[9] Barzegar A, Ghadipasha M, Rezaei N, Forouzesh M, Valizadeh R. New hope for treatment of respiratory involvement following COVID-19 by bromhexine. Journal of Nephropharmacology. 2021;**10**:e11. DOI: 10.34172/npj.2021.11

[10] Ghiasi N, Valizadeh R, Arabsorkhi M, Hoseyni TS, Esfandiari K, Sadighpour T, et al. Efficacy and side effects of Sputnik V, Sinopharm and AstraZeneca vaccines to stop COVID-19: A review and discussion. Immunopathologia Persa. 2021;**7**:e31. DOI: 10.34172/ipp.2021.31

[11] Shah K, Saxena D, Mavalankar D. Vitamin D supplementation, COVID-19 and disease severity: A meta-analysis. QJM. 2021;**114**:175-181. DOI: 10.1093/qjmed/hcab009

[12] Garland CF, Garland FC, Gorham ED, Lipkin M, Newmark H, Mohr SB, et al. The role of vitamin D in cancer prevention. American Journal of

Public Health. 2006;**96**:252-261. DOI: 10.2105/AJPH.2004.045260

[13] Zittermann A. Vitamin D in preventive medicine: Are we ignoring the evidence? The British Journal of Nutrition. 2003;**89**:552-572. DOI: 10.1079/BJN2003837

[14] Cao X. COVID-19: Immunopathology and its implications for therapy. Nature Reviews. Immunology. 2020;**20**:269-270. DOI: 10.1038/s41577-020-0308-3

[15] Subramaniam M, Chong S-A, Pek E. Diabetes mellitus and impaired glucose tolerance in patients with schizophrenia. The Canadian Journal of Psychiatry. 2003;**48**:345-347. DOI: 10.1177/070674370304800512

[16] Suvisaari J, Mantere O, Keinänen J, Mäntylä T, Rikandi E, Lindgren M, et al. Is it possible to predict the future in first-episode psychosis? Frontiers in Psychiatry. 2018;**9**:580. DOI: 10.3389/fpsyt.2018.00580

[17] Zareifopoulos N, Bellou A, Spiropoulou A, Spiropoulos K. Prevalence of comorbid chronic obstructive pulmonary disease in individuals suffering from schizophrenia and bipolar disorder: A systematic review. COPD: Journal of Chronic Obstructive Pulmonary Disease. 2018;**15**:612-620. DOI: 10.1080/15412555.2019.1572730

[18] Hjorthøj C, Stürup AE, McGrath JJ, Nordentoft M. Years of potential life lost and life expectancy in schizophrenia: A systematic review and meta-analysis. The Lancet Psychiatry. 2017;**4**:295-301. DOI: 10.1016/S2215-0366(17)30078-0

[19] Green AI, Canuso CM, Brenner MJ, Wojcik JD. Detection and management of comorbidity in patients with schizophrenia. Psychiatric Clinics. 2003;**26**:115-139. DOI: 10.1016/s0193-953x(02)00014-x

[20] Olfson M, Gerhard T, Huang C, Crystal S, Stroup TS. Premature mortality among adults with schizophrenia in the United States. JAMA Psychiatry. 2015;**72**:1172-1181. DOI: 10.1001/jamapsychiatry.2015.1737

[21] Viani-Walsh D, Kennedy-Williams S, Taylor D, Gaughran F, Lally J. Vitamin D deficiency in schizophrenia implications for COVID-19 infection. Irish Journal of Psychological Medicine. 2021;**38**:278-287. DOI: 10.1017/ipm.2020.107

[22] Eyles DW, Trzaskowski M, Vinkhuyzen AA, Mattheisen M, Meier S, Gooch H, et al. The association between neonatal vitamin D status and risk of schizophrenia. Scientific Reports. 2018;**8**:1-8. DOI: 10.1038/s41598-018-35418-z

[23] Fond G, Pauly V, Orleans V, Antonini F, Fabre C, Sanz M, et al. Increased in-hospital mortality from COVID-19 in patients with schizophrenia. L'Encéphale. 2021;**47**:89-95. DOI: 0.1016/j.encep.2020.07.003

[24] Bradford DW, Kim MM, Braxton LE, Marx CE, Butterfield M, Elbogen EB. Access to medical care among persons with psychotic and major affective disorders. Psychiatric Services. 2008;**59**:847-852. DOI: 10.1176/ps.2008.59.8.847

[25] Lawrence D, Kisely S. Inequalities in healthcare provision for people with severe mental illness. Journal of Psychopharmacology. 2010;**24**:61-68. DOI: 10.1177/1359786810382058

[26] Zhu J-l, Luo W-w, Cheng X, Li Y, Zhang Q-z, Peng W-x. Vitamin D deficiency and schizophrenia in adults: A systematic review and meta-analysis

of observational studies. Psychiatry Research. 2020;**288**:112959. DOI: 10.1016/j.psychres.2020.112959

[27] Torrey EF, Miller J, Rawlings R, Yolken RH. Seasonality of births in schizophrenia and bipolar disorder: A review of the literature. Schizophrenia Research. 1997;**28**:1-38. DOI: 10.1016/s0920-9964(97)00092-3

[28] McGrath J, Scott J. Urban birth and risk of schizophrenia: A worrying example of epidemiology where the data are stronger than the hypotheses. Epidemiology and Psychiatric Sciences. 2006;**15**:243-246

[29] Marcelis M, Takei N, van Os J. Urbanization and risk for schizophrenia: Does the effect operate before or around the time of illness onset? Psychological Medicine. 1999;**29**:1197-1203. DOI: 10.1017/s0033291799008983

[30] Mortensen PB, Pedersen CB, Westergaard T, Wohlfahrt J, Ewald H, Mors O, et al. Effects of family history and place and season of birth on the risk of schizophrenia. The New England Journal of Medicine. 1999;**340**:603-608. DOI: 10.1056/NEJM199902253400803

[31] March D, Hatch SL, Morgan C, Kirkbride JB, Bresnahan M, Fearon P, et al. Psychosis and place. Epidemiologic Reviews. 2008;**30**:84-100. DOI: 10.1093/epirev/mxn006

[32] Saha S, Chant D, Mcgrath J. Meta-analyses of the incidence and prevalence of schizophrenia: Conceptual and methodological issues. International Journal of Methods in Psychiatric Research. 2008;**17**:55-61. DOI: 10.1002/mpr.240

[33] Hintzpeter B, Scheidt-Nave C, Müller MJ, Schenk L, Mensink GB. Higher prevalence of vitamin D deficiency is associated with immigrant background among children and adolescents in Germany. The Journal of Nutrition. 2008;**138**:1482-1490. DOI: 10.1093/jn/138.8.1482

[34] Del Valle HB, Yaktine AL, Taylor CL, Ross AC. Dietary Reference Intakes for Calcium and Vitamin D. Washington (DC): National Academies Press; 2011. DOI: 10.17226/13050

[35] Wagner CL, McNeil R, Hamilton SA, Winkler J, Cook CR, Warner G, et al. A randomized trial of vitamin D supplementation in 2 community health center networks in South Carolina. American Journal of Obstetrics and Gynecology. 2013;**208**:137.e1-137.e13. DOI: 10.1016/j.ajog.2012.10.888

[36] McGrath J, Saari K, Hakko H, Jokelainen J, Jones P, Järvelin M-R, et al. Vitamin D supplementation during the first year of life and risk of schizophrenia: A Finnish birth cohort study. Schizophrenia Research. 2004;**67**:237-245. DOI: 10.1016/j.schres.2003.08.005

[37] Cashman KD. Global differences in vitamin D status and dietary intake: A review of the data. Endocrine Connections. 2022;**11**(1):e210282. DOI: 10.1530/EC-21-0282

[38] Hilger J, Friedel A, Herr R, Rausch T, Roos F, Wahl DA, et al. A systematic review of vitamin D status in populations worldwide. The British Journal of Nutrition. 2014;**111**:23-45. DOI: 10.1017/S0007114513001840

[39] Holick MF, Chen TC. Vitamin D deficiency: A worldwide problem with health consequences. The American Journal of Clinical Nutrition. 2008;**87**:1080S-1086S. DOI: 10.1093/ajcn/87.4.1080S

[40] Ross AC, Manson JE, Abrams SA, Aloia JF, Brannon PM, Clinton SK, et

al. The 2011 report on dietary reference intakes for calcium and vitamin D from the Institute of Medicine: What clinicians need to know. The Journal of Clinical Endocrinology and Metabolism. 2011;**96**:53-58. DOI: 10.1210/jc.2010-2704

[41] Valipour G, Saneei P, Esmaillzadeh A. Serum vitamin D levels in relation to schizophrenia: A systematic review and meta-analysis of observational studies. The Journal of Clinical Endocrinology and Metabolism. 2014;**99**:3863-3872. DOI: 10.1210/jc.2014-1887

[42] Cui X, McGrath JJ, Burne TH, Eyles DW. Vitamin D and schizophrenia: 20 years on. Molecular Psychiatry. 2021;**26**:2708-2720. DOI: 10.1038/s41380-021-01025-0

[43] Sheikhmoonesi F, Zarghami M, Mamashli S, Charati JY, Hamzehpour R, Fattahi S, et al. Effectiveness of vitamin D supplement therapy in chronic stable schizophrenic male patients: A randomized controlled trial. Iranian Journal of Pharmaceutical Research. 2016;**15**:941

[44] Tiangga E, Gowda A, Dent JA. Vitamin D deficiency in psychiatric in-patients and treatment with daily supplements of calcium and ergocalciferol. Psychiatric Bulletin. 2008;**32**:390-393. DOI: 10.1192/pb.bp.107.019109

[45] Ali N. Role of vitamin D in preventing of COVID-19 infection, progression and severity. Journal of Infection and Public Health. 2020;**13**:1373-1380. DOI: 10.1016/j.jiph.2020.06.021

[46] Speeckaert MM, Delanghe JR. Association between low vitamin D and COVID-19: Don't forget the vitamin D binding protein. Aging Clinical and Experimental Research. 2020;**32**:1207-1208. DOI: 10.1007/s40520-020-01607-y

[47] Akbar MR, Wibowo A, Pranata R, Setiabudiawan B. Low serum 25-hydroxyvitamin D (vitamin D) level is associated with susceptibility to COVID-19, severity, and mortality: A systematic review and meta-analysis. Frontiers in Nutrition. 2021;**8**:131. DOI: 10.3389/fnut.2021.660420

[48] Rejnmark L, Bislev LS, Cashman KD, Eiríksdottir G, Gaksch M, Grübler M, et al. Non-skeletal health effects of vitamin D supplementation: A systematic review on findings from meta-analyses summarizing trial data. PLoS One. 2017;**12**:e0180512. DOI: 10.1371/journal.pone.0180512

[49] Pereira M, Dantas Damascena A, Galvão Azevedo LM, de Almeida OT, da Mota SJ. Vitamin D deficiency aggravates COVID-19: Systematic review and meta-analysis. Critical Reviews in Food Science and Nutrition. 2022;**62**:1308-1316. DOI: 10.1080/10408398.2020.1841090

[50] Zhou Y-F, Luo B-A, Qin L-L. The association between vitamin D deficiency and community-acquired pneumonia: A meta-analysis of observational studies. Medicine. 2019;**98**(38):e17252. DOI: 10.1097/MD.0000000000017252

[51] Adami S, Bertoldo F, Braga V, Fracassi E, Gatti D, Gandolini G, et al. 25-hydroxy vitamin D levels in healthy premenopausal women: Association with bone turnover markers and bone mineral density. Bone. 2009;**45**:423-426. DOI: 10.1016/j.bone.2009.05.012

[52] Pimenta FB, Pinho L, Silveira MF, Botelho ACC. Factors associated with chronic diseases among the elderly receiving treatment under the Family Health Strategy. Ciência & Saúde

Coletiva. 2015;**20**:2489-2498. DOI: 10.1590/1413-81232015208.11742014

[53] Chen N, Wan Z, Han S-F, Li B-Y, Zhang Z-L, Qin L-Q. Effect of vitamin D supplementation on the level of circulating high-sensitivity C-reactive protein: A meta-analysis of randomized controlled trials. Nutrients. 2014;**6**:2206-2216. DOI: 10.3390/nu6062206

[54] Austad SN, Fischer KE. Sex differences in lifespan. Cell Metabolism. 2016;**23**:1022-1033. DOI: 10.1016/j.cmet.2016.05.019

[55] Ahrenfeldt LJ, Otavova M, Christensen K, Lindahl-Jacobsen R. Sex and age differences in COVID-19 mortality in Europe. Wiener Klinische Wochenschrift. 2021;**133**:393-398. DOI: 10.1007/s00508-020-01793-9

[56] Raoult D, Zumla A, Locatelli F, Ippolito G, Kroemer G. Coronavirus infections: Epidemiological, clinical and immunological features and hypotheses. Cell Stress. 2020;**4**:66. DOI: 10.15698/cst2020.04.216

[57] Sánchez-PernauteA,Pérez-AguirreME, Jiménez AP, Campos AR, Muñoz A, Torres A. Intraluminal mesh erosion after prosthetic hiatoplasty: Incidence, management, and outcomes. Diseases of the Esophagus. 2019;**32**. DOI: 10.1093/dote/doy131

[58] Yanez ND, Weiss NS, Romand J-A, Treggiari MM. COVID-19 mortality risk for older men and women. BMC Public Health. 2020;**20**:1-7. DOI: 10.1186/s12889-020-09826-8

[59] Ye R, Huang J, Wang Z, Chen Y. Trace element selenium effectively alleviates intestinal diseases. International Journal of Molecular Sciences. 2021;**22**(21):11708. DOI: 10.3390/ijms222111708

[60] Williamson EJ, Walker AJ, Bhaskaran K, Bacon S, Bates C, Morton CE, et al. Factors associated with COVID-19-related death using OpenSAFELY. Nature. 2020;**584**:430-436. DOI: 10.1038/s41586-020-2521-4

[61] Wang Q, Xu R, Volkow ND. Increased risk of COVID-19 infection and mortality in people with mental disorders: Analysis from electronic health records in the United States. World Psychiatry. 2021;**20**:124-130. DOI: 10.1002/wps.20806

[62] Correll CU, Solmi M, Veronese N, Bortolato B, Rosson S, Santonastaso P, et al. Prevalence, incidence and mortality from cardiovascular disease in patients with pooled and specific severe mental illness: A large-scale meta-analysis of 3,211,768 patients and 113,383,368 controls. World Psychiatry. 2017;**16**:163-180. DOI: 10.1002/wps.20420

[63] Vancampfort D, Correll CU, Galling B, Probst M, De Hert M, Ward PB, et al. Diabetes mellitus in people with schizophrenia, bipolar disorder and major depressive disorder: A systematic review and large scale meta-analysis. World Psychiatry. 2016;**15**:166-174. DOI: 10.1002/wps.20309

[64] Vancampfort D, Wampers M, Mitchell AJ, Correll CU, De Herdt A, Probst M, et al. A meta-analysis of cardio-metabolic abnormalities in drug naïve, first-episode and multi-episode patients with schizophrenia versus general population controls. World Psychiatry. 2013;**12**:240-250. DOI: 10.1002/wps.20069

[65] HLA-C H. Common polygenic variation contributes to risk of schizophrenia and bipolar disorder. Nature. 2009;**460**:748-752. DOI: 10.1038/nature08185

[66] Fond G, Nemani K, Etchecopar-Etchart D, Loundou A, Goff DC, Lee SW,

et al. Association between mental health disorders and mortality among patients with COVID-19 in 7 countries: A systematic review and meta-analysis. JAMA Psychiatry. 2021;**78**:1208-1217. DOI: 10.1001/jamapsychiatry.2021.2274

[67] Charlson FJ, Ferrari AJ, Santomauro DF, Diminic S, Stockings E, Scott JG, et al. Global epidemiology and burden of schizophrenia: Findings from the global burden of disease study 2016. Schizophrenia Bulletin. 2018;**44**:1195-1203. DOI: 10.1093/schbul/sby058

[68] Ćurić M, Zafirovski O, Spiridonov V. Essentials of Medical Meteorology. Cham: Springer; 2022. DOI: 10.1007/978-3-030-80975-1

[69] Shelef A, Dahan S, Weizman S. Vitamin D as a protective factor in COVID-19 infection in elderly schizophrenia and dementia inpatients: A case series. The Israel Medical Association Journal. 2022;**24**:74-77

Chapter 7

Vitamin D Deficiency: Implications in COVID-19 and Schizophrenia

Sepehr Saberian, Fahim Atif, Donald Stein and Seema Yousuf

Abstract

Deficiencies in vitamin D can have several etiologies, broadly classified as the following: suboptimal exposure to ultraviolet-B (UV-B) light from sunlight, low dietary intake of vitamin-D or reduced absorption due to gastrointestinal pathologies, reduced production due to liver or kidney disease, pseudo-deficiencies caused by end organ resistance despite normal or elevated vitamin D levels, and medication-induced stimulation of hepatic cytochrome P450 enzymes for which vitamin D is a substrate. Deficiencies in this important vitamin can have several adverse clinical implications such as osteomalacia, osteoporosis, muscle pain, and depression to name a few. More recently, vitamin D has been shown to be involved in modulating various aspects of the immune system. Vitamin D receptors have also been found to be present in certain regions of the brain, especially those involved in schizophrenia. We will discuss the implications of vitamin D deficiency and its immunomodulatory role in the setting of the COVID-19 virus, the proposed cellular and molecular mechanisms of action for vitamin D in the context of schizophrenia, and the clinical outcomes associated with these two pathologies as a function of low vitamin D levels.

Keywords: vitamin D deficiency, schizophrenia, COVID-19, immunomodulation

1. Introduction

Vitamin D plays a crucial role in several biologic processes. As such, maintaining physiologic levels of this vitamin is essential for the proper functioning of various organ systems. Unfortunately, vitamin D deficiency (VDD) is a global health concern with potentially severe clinical outcomes [1]. This is true for both adults as well as the pediatric population. Although a wide range of risk factors exists, the most well-studied and accepted risk factor remains lack of sun exposure. Certain studies have also looked at race and skin color as potential risk factors (more on this later) [2]. Risk factor associated with VDD is an area of continuing research.

Although bone disease is the most well-established consequence of VDD, it is important to appreciate the complex and nuanced ways in which the endocrine system, intestines, and kidneys interact with and depend on vitamin D [3]. Furthermore, previously unknown functionalities of this vitamin have been elucidated in recent years. Perhaps the most fascinating findings have been vitamin D's ability to modulate the immune system. Various experiments have shown the surprising ability of vitamin D to stimulate the immune system to mount a more potent and effective response against foreign pathogens. Interestingly, these observations have been made in both the innate and adaptive immune systems [4]. Another area of research in the setting of vitamin D has been the brain [5]. Detection of vitamin D receptors in the central nervous system has provided an avenue for researchers to examine how the vitamin may be implicated in various diseases as well as the developmental stages of the brain.

The significant amount of information on these new areas, as well as the excitement surrounding it, is evident in the literature. We have analyzed several studies and experiments to provide an informative and structured review of the most recent progress and discoveries in this area. In this chapter, we first approach VDD broadly by discussing its epidemiology, pathogenesis, and pathophysiology. We then narrow the scope of our discussion to focus on the most recent and exciting findings in the context of vitamin D. Finally, we conclude our review of VDD by exploring how these recent findings are implicated in schizophrenia as well as the novel COVID-19 virus at much more granular level.

2. Epidemiology

The diagnosis of VDD is made when levels of 25-hyroxy vitamin D, or 25(OH)D, levels are found to be below the threshold value of 12ng/mL [3]. The global prevalence of VDD is estimated to be 14-59% in the adult population. Although there is a paucity of data on infants and children in several countries, the global prevalence of VDD in this group is estimated to be higher than the adult population, with some studies reporting rates as high as 80% [6–8]. Furthermore, prevalence of the disease varies significantly by region, age group, season in which measurements were taken, and gender. Of note is the markedly higher rates seen in the Middle East, especially in Iran(infants, 86%; adults, 51%) [7, 9]. VDD is also seen in some South Asian countries such as India, Pakistan, and Bangladesh, where ~80% of adults are known to be affected; the same statistic in US adults has been reported as ~35% [1]. Conversely, the European population has a relatively lower prevalence compared to other regions, with an estimated 8.3% - 17.7% of the population affected [10].

A significant risk factor for VDD is race. For instance, in the US, 82.1% of African American adults and 62.9% of Hispanic adults are deficient in vitamin D; this is in comparison to the overall VDD rate of 41.6%. This is attributable to the relatively higher melanin levels observed in the skin of these individuals [11]. Melanin is a polymer that not only provides skin pigmentation, but also absorbs ultraviolet (UV) radiation [2, 12]. By doing so, melanin decreases the amount of UV light available to keratinocytes (which are a crucial component of vitamin D synthesis) located within the skin epidermis [13].

In recent decades, younger individuals have been at much higher risk of VDD. This is largely thought to be a result of the accelerating use of technology. Electronic devices such as video game consoles, portable tablets, computers, and cell phones have provided children with entertainment that can be enjoyed indoors. As a result, they are less likely to engage in outdoor activities, which has significantly decreased exposure to sunlight in this age group [11, 14]. Although obesity has also been reported as a risk factor for VDD with various proposed mechanisms of action, a portion of the VDD observed in obese individuals may be a result of confounding effects [15]. In other words, it is possible that obese children also tend to spend more time indoors and do not get sufficient sun exposure. Further studies investigating the true effect of obesity on VDD would require comparing data on obese children who engage in outdoors activities to obese children who do not engage in such activities.

3. Pathogenesis

There are several reasons why VDD may develop. It can be helpful to broadly categorize these etiologies. Although these are varied in their mechanisms of action, they all lead to absolute or functional deficiencies in vitamin D:

- Suboptimal exposure to ultraviolet-B (UV-B) light from sunlight: The precursor molecules to initiating vitamin D synthesis, located in the skin, require exposure to UV-B light. As a result, low levels of sun exposure cause decreased vitamin D synthesis [1].

- Low dietary intake of vitamin D: The vitamin can also be absorbed through the intestinal tract. Consuming foods that do not contain adequate vitamin D can result in VDD.

- Reduced absorption due to gastrointestinal pathologies: Certain diseases of the gastrointestinal tract can result in decreased ability of the intestinal lining to absorb various nutrients, including vitamin D [16].

- Reduced production due to liver or kidney disease: As we will discuss later, the liver and kidneys contain important enzymes that are required to produce functional vitamin D. Chronic liver or kidney disease can result in sub-physiologic levels of these enzymes, leading to decreased vitamin D production [17, 18].

- Pseudo-deficiencies caused by end organ resistance despite normal or elevated vitamin D levels: Because vitamin D exerts its effects by binding to the vitamin D receptor, signs and symptoms of vitamin D deficiency can manifest if the receptor does not function properly. As a result, vitamin D levels can be normal or even elevated [19].

- Medication-induced stimulation of hepatic cytochrome P450 enzymes: The cytochrome P450 enzymes are a family of proteins found in the liver that are responsible for metabolizing a wide range of substrates. Specific cytochrome P450 enzymes, for which vitamin D is a substrate, can be stimulated as a side effect of certain medications. This in turn leads to hypermetabolism of vitamin D and with chronic use, can lead to VDD [20].

4. Pathophysiology

Examination of the pathophysiology of VDD and how it interacts with various organs can clarify the various roles this vitamin plays. Vitamin D3 and vitamin D2, precursors to mature vitamin D, must first enter the liver where they are converted to 25(OH)D [21]. Following another enzymatic conversion by the kidneys, 25(OH)D becomes 1,25(OH)D; this is the mature and functional form of vitamin D [22].

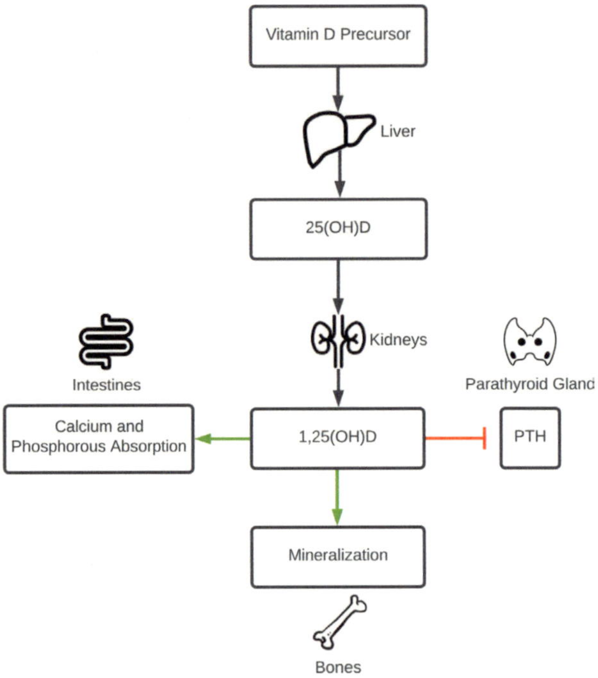

Figure 1.
Vitamin D Physiology: Classical Pathway. The Vitamin D precursor enters the liver, where it is enzymatically modified to yield 25(OH)D. It then enters the kidneys, where it is once again enzymatically modified to yield 1,25(OH)D, the functional form of vitamin D. In the intestines, calcium and phosphorous absorption is stimulated (green arrow); in the bones, mineralization is stimulated (green arrow); in the parathyroid glands, PTH (parathyroid hormone) secretion is inhibited (red arrow).

Classically, there are three pathways in which vitamin D has been implicated: the endocrine system by way of the parathyroid glands, the gastrointestinal tract, and the skeletal system. Vitamin D inhibits the secretion of parathyroid hormone (PTH) by the parathyroid glands (PTH increases serum calcium levels, decreases phosphorous levels, and stimulates bone resorption), stimulated intestinal absorption of calcium and phosphate, and increased mineralization in the bones [23]. These pathways are outlined in **Figure 1**.

Having discussed the normal functions of vitamin D, we will now examine the sequalae of VDD. Low levels of vitamin D stimulate PTH release from the parathyroid glands, leading to increased serum calcium levels and decreased phosphorous levels. Elevated PTH also induces bone resorption (or breakdown). There is also decreased stimulation of the intestine for absorption of calcium and phosphorous as well as decreased bone mineralization in the setting of VDD. The overall result of VDD, therefore, is decreased bone density, decreased serum phosphorous and calcium levels, and elevated PTH [1]. Depending on the severity of the deficiency, symptomatology can be nuanced and may include any combination of the list included below [24].

- Bone fractures from mild trauma
- Bone pain
- Joint pain

- Muscle pain
- Fatiguability
- Psychological symptoms

This section has focused solely on the classical roles of vitamin D and the clinical outcomes observed as a result of VDD. In addition to these classical roles, there are other roles of vitamin D that have been elucidated more recently. These will be discussed in detail in the following section.

5. Recent developments: schizophrenia and COVID-19

This section will examine new insights on the roles of vitamin D. The main topics of discussion will center around the implications of vitamin D in the immune system as well as the central nervous system. Furthermore, we will present a more focused discussion of these topics in the context of schizophrenia and COVID-19.

5.1 Schizophrenia

Schizophrenia is an often debilitating, chronic mental disorder, where patients present with symptoms such as hallucinations, delusions, altered perception, as well as disorganized speech and behavior [25]. It has been well established that an imbalance of various neurotransmitters in the brain is responsible for these symptoms. The most widely studies neurotransmitters in the context of schizophrenia include dopamine, serotonin, and glutamate [26]. Of these, dopamine activity appears to have the strongest correlation to symptomatology by modulating dopamine-1 (D1) and dopamine-2 (D2) receptors [27, 28]. Specifically, dopamine's effects on four main pathways within the brain have been implicated in schizophrenia and medication side effects: the mesocortical pathway, mesolimbic pathway, nigrostriatal pathway, and tuberoinfundibular pathway. In terms of disease symptoms, the combination of decreased activity of the mesocortical pathway and increased activity of the mesolimbic pathways are the culprit.

On the other hand, the nigrostriatal and tuberoinfundibular pathways are only affected when anti-psychotic medications are administered [29]. This is due to the presence of D2 receptors in all four regions. Because schizophrenia is a disease of localized dopamine dysregulation in the brain, pharmacologic intervention requires blockage of D2 receptors to restore homeostatic dopamine activity. However, these medications do not specifically target the mesocortical and mesolimbic pathways, but instead act on all brain regions that contain D2 receptors. As mentioned, the D2 receptor is also present in the nigrostriatal and tuberoinfundibular pathways, with unintended blockage causing undesirable side effects including movement disorders (extrapyramidal effects), milk discharge (galactorrhea), enlarged breasts, sexual dysfunction, among others [30].

Most recently, vitamin D's role in schizophrenia has become major a topic of interest for many researchers. Not only has VDD been shown to be associated with schizophrenia, but it has also been implicated in individuals experiencing single episodes of psychosis [31]. In one report, vitamin D levels of three study groups consisting of schizophrenic patients in remission, schizophrenic patients having an acute episode, and patients without psychiatric illness were analyzed and compared. Interestingly, patients who were in the midst of an acute episode were found to have significantly lower vitamin D levels, with a median of 7.18 ng/mL, as compared to both patients in remission (15.03 ng/mL) and non-schizophrenic patients (15.02 ng/mL) [32]. The authors concluded that there exists a clear association between low vitamin D levels and schizophrenic episodes. This does not imply a causative effect, however, as interactions with other pathways must be considered. In fact, a 2015 study searching for potential interactions of vitamin D with other pathways reported proline as one candidate. It was found that the proline dehydrogenase, or *PRODH*, gene's transcription was significantly modulated by vitamin D [33]. This enzyme plays an important role in proline catabolism; as a result, proline levels decreased with increasing proline dehydrogenase concentrations and vice versa [34]. Not surprisingly, VDD was found to be associated with higher proline levels. Furthermore, hyperprolinemia was shown to contribute to 33% of the relationship between VDD and schizophrenia [33].

Recently, the presence of vitamin D receptors (VDR) in the brain has been confirmed. More specifically, high concentrations of this receptor have been demonstrated in the hippocampus, supraoptic and paraventricular nuclei, and substantia nigra [35]. Interestingly, organs classically associated with the site of vitamin D activity (such the kidneys, bone, and gut) contain multiple VDR isoforms, however, the brain contains only one isoform [36]. Once the vitamin D-VDR complex forms, it induces various downstream pathways by binding DNA response elements [5]. These VDR can be found in both the adult and the developing brain. As such, VDD in the developing brain can have serious implications. Studies utilizing rodent models have shown that subphysiologic levels of vitamin D in offspring led to abnormalities in neuronal differentiation, altered anatomy, neurotransmitter imbalance, and abnormal gene expression [37]. These findings are also accompanied by abnormal behavioral and cognitive observations, further confirming the vital role of vitamin D in the proper development of the brain [37, 38]. Two retrospective human studies have also been conducted, both of which demonstrated a significant association between neonatal VDD and increased risk of later developing schizophrenia in adulthood [36].

Although it can be useful to understand the mechanisms by which VDD affects the brain and potentially causes schizophrenia, it is even more important to examine whether vitamin D repletion can restore normal brain function. In one study, administration of vitamin D for an eight-week period in schizophrenic patients treated with the medication Clozapine has been shown to improve cognition, without significant effects on psychotic episodes [39]. Interestingly, a similar study demonstrated that in addition to improving cognition, vitamin D supplementation led to improved symptoms in patients suffering from schizophrenia. Of note is that in this study, vitamin D supplementation was not constrained by the wight-week period, but instead vitamin D levels of > 30ng/mL was used as the threshold [40]. This is an important consideration, because it provides an explanation as to why one study reported isolated improvement in cognition while the other reported improved cognition as well as psychotic symptoms. These results imply that there exists a concentration-dependent relationship between vitamin D levels and symptom improvement.

Considering this body of evidence that has recently become available, it is clear that vitamin D is a vital component for not only proper functioning of the adult brain, but also appropriate growth and maturing of the developing brain. We have also discussed insights gathered from recent studies on the link between VDD and schizophrenia. Lastly, it is important to consider the various levels at which we have examined this topic. From a basic science standpoint, we discussed microscopic and macroscopic changes resulting from VDD seen in the neonatal brain and potential cellular and extracellular mediators of disease. From a clinical standpoint, we've reviewed several studies that explored the relationship between VDD and the risk of developing schizophrenia and lastly, we discussed studies in which vitamin D supplementation was shown to improve schizophrenia symptoms. Next, we will explore associations between VDD and the COVID-19 virus, how it might affect various aspects of the disease, and whether supplementation with vitamin D has been shown to be beneficial in mitigating the severity of the disease.

5.2 COVID-19

The coronavirus disease 2019 (COVID-19), which first began as an endemic, but rapidly spread to become a global pandemic, has affected millions of people. With death tolls rising at an unprecedented rate, medical research set out to identify potential therapeutic and prophylactic agents to battle the COVID-19 virus and accompanying fatal symptoms [41]. The main syndrome associated with the COVID-19 is acute respiratory distress syndrome (ARDS). ARDS results from overactivation of the immune system, causing extravasation of large volume of fluids into the lungs [42]. This syndrome is extremely dangerous and can lead to death even with maximal medical intervention. In the search for answers to a therapeutic solution, one viable candidate has been vitamin D.

As previously discussed, recent interest in the role of vitamin D in various physiologic processes has provided a plethora of new insights. One such finding has been vitamin D's role in modulating the body's immune system. The mechanism by which this is accomplished involves the same VDR mentioned in the previous section. The receptor has been found in a vast number of immune cells and is thought to modulate the immune system in this way. High concentrations have been found particularly in antigen-presenting cells (APC) such as dendritic cells, $CD4^+$ and $CD8^+$ lymphocytes, as well as macrophages [43]. As an intracellular receptor, VDR binds with vitamin D and upon activation, can regulate the transcription of a number of genes [44]. Transcriptional regulation is not however the only way that vitamin D exerts its effects. It has also been shown to interact directly with protein other than VDR, modify histone and chromatin structure, among others [45]. To understand how vitamin D relates to the COVID-19 virus, we will first examine the immune system's role, then we will explore how these findings related to COVID-19 specifically.

The immune system has one basic, yet functionally complex goal: to destroy foreign particles that pose a threat to the body. The intricate network of immune cells works to neutralize such threats. In order to do so, they communicate with one and another by way of cytokines and other hormones, which are chemicals that act as messengers by binding to their intended receptors located on cell surfaces. Within the innate immune system, signaling via IFN-γ, STAT-1α, lipopolysaccharide (LPS) and toll-like receptors (TLRs) has been shown to increase the activity of 1α- hydroxylase levels in monocytes; this enzyme is responsible for the last step of vitamin D synthesis [4]. Activation of the enzyme, and the ensuing surge of vitamin D levels in monocytes, leads to differentiation of these cells into mature macrophages. Interestingly, in both macrophages and dendritic cells, vitamin D induces an anti-inflammatory state by simultaneously decreasing pro-inflammatory and increasing anti-inflammatory cytokines [46]. Additionally, lymphocytes also express fewer inflammatory receptors in response to vitamin D.

Within the adaptive immune system, the effects of vitamin D can be more nuanced. For instance, vitamin D induces apoptosis of activated B cells and decreases production of plasma cells. Importantly, there is no effect on B cell differentiation. T cells may respond differentially to vitamin D as a function of cellular state and phenotype [43]. The vitamin D-T cell interaction can result in the downregulating the levels of several cytokines including IL-2, IFN-γ, IL-17, and IL-21; the end result is an overall anti-inflammatory state [46]. One important observation that is common to both B and T cells is the markedly decreased proliferation of autoreactive cells [47]. These cells are responsible for various autoimmune disorders, and as multiple studies have shown, vitamin D can be therapeutic in this setting. Given vitamin D's extensive immunomodulatory role, it has become increasingly evident that it may have major implications in infectious diseases. In the context of COVID-19, the viral pathogen's main entry point into the body is the respiratory system. Not only is there a high concentration of VDR in the macrophages located in the lung epithelium, but there are also high levels within the epithelial cells themselves. Activation of VDR in the epithelium stimulates production of several anti-microbial proteins that hinder the entrance of such pathogens. Concurrently, VDR activation in lung macrophages is thought to prevent immune system hyperactivity by way of decreasing inflammatory signals [48].

Vitamin D does not only modulate the effects of COVID-19 by way of immunomodulation. Another very important physiologic pathway involved in COVID-19 infection is the renin-angiotensin-aldosterone system (RAAS). RAAS is involved in regulating blood volume and pressure. It's imperative to understand how this system works physiologically in order to understand how COVID-19 causes its pathological sequalae. Renin is a hormone produced and secreted by the kidneys in response to changes in blood volume or blood pressure. If either of these parameters are decreased, renin is secreted and induces the conversion of angiotensinogen (produced by the liver) to angiotensin I (AT1). Subsequently, angiotensin I travels to the lungs through the systemic circulation. The lungs contain the enzyme angiotensin converting enzyme (ACE), which is responsible for converting AT1 to angiotensin II (AT2). AT2 then acts on several end organs, resulting in increased blood pressure and volume. The enzyme responsible for breaking down AT2 is angiotensin converting enzyme 2(ACE2). In the setting of COVID-19 infection, the virus causes abnormal downregulation of ACE2. This results in the inability to degrade AT2, yielding exceptionally high concentrations of the hormone. Increased AT2 levels then activate the RAAS, which cause hypertension and above physiologic blood volume. With this dramatic increase in blood volume and pressure, substantially more fluid leaks into the lung parenchyma causing ARDS. Vitamin D has been shown to act at several levels to mitigate these pathologic processes. At the level of RAAS biosynthesis, vitamin D acts as a negative regulator by inhibiting renin synthesis [49]. It also has been shown to increase ACE2 levels, allowing more AT-2 breakdown [50]. Lastly, vitamin D has downstream vasodilatory effects which provide a counterforce against the vasoconstriction caused by AT-2 [51, 52].

Taking together the immunomodulatory functions of vitamin D as well as its role in regulating the RAAS system, we can understand its significant therapeutic potential in the setting of COVID-19. In the immune system, vitamin D stimulates a more robust immune response against pathogens, such as the COVID-19 virus; by modulating RAAS, it decreases volume overload in the circulatory system and potentially decreases the likelihood of developing ARDS.

6. Conclusions

In this chapter, we have discussed several key topics related to VDD. These included the epidemiology, recent basic science and clinical developments, and the role that vitamin D plays in schizophrenia and COVID-19. Epidemiologically, there is a staggering portion of the population who suffers from VDD; this is especially true of certain countries in the Middle East. Aside from geographic location, darker skin colors are also significantly associated with higher rates of VDD. There are also several additional modifiable and non-modifiable risk factors associated with suboptimal vitamin D levels, as discussed previously.

From a basic science standpoint, there have been a number of new discoveries and developments in identifying the role of vitamin D in organ systems besides those involved in the classical pathways. These include the vitamin's role in modulating the immune system, regulating the circulatory system by way of RAAS, the proper functioning of the central nervous system, as well as appropriate development of the growing brain. Implications of these new findings can then be analyzed in the clinical setting. In the context of schizophrenia, vitamin D supplementation, in a dose and concentration-dependent manner, has been shown to improve symptoms. This has been attributed to the detection of vitamin D receptors in the brain.

Furthermore, the immunologic and circulatory regulation capabilities of vitamin D have made it a topic of interest in searching for a treatment for COVID-19 infection. By stimulating various immune cells involved in both the innate and adaptive immune system, vitamin D plays a role in neutralizing and clearing the COVID-19 virus. Additionally, a feared consequence of the infection is ARDS. By counteracting the pathological disruptions in the RAAS system, vitamin D may help decrease the severity of or even prevent ARDS.

In conclusion, we have seen that vitamin D's functions in the body are far more nuanced than previously thought. The new insights discussed in this chapter provide a broader range of both physiologic and pathophysiologic effects that this crucial vitamin has throughout the body. Although the roles of vitamin D in the endocrine, gastrointestinal, and skeletal systems are extremely important, it is imperative to also consider its immune system, circulatory system, and nervous system implications moving forward.

Conflict of interest

The authors declare no conflict of interest.

Author details

Sepehr Saberian[1], Fahim Atif[2], Donald Stein[2] and Seema Yousuf[2*]

1 Morehouse School of Medicine, Atlanta, USA

2 Emory University School of Medicine, Atlanta, USA

*Address all correspondence to: seema.yousuf@emory.edu

References

[1] Sizar O, Khare S, Goyal A, Givler A. Vitamin D Deficiency. Treasure Island (FL): StatPearls Publishing; 2022

[2] Webb AR, Kazantzidis A, Kift RC, Farrar MD, Wilkinson J, Rhodes LE. Colour counts: Sunlight and skin type as drivers of Vitamin D deficiency at UK latitudes. Nutrients. 2018;**10**(4):457

[3] Amrein K, Scherkl M, Hoffmann M, Neuwersch-Sommeregger S, Köstenberger M, Tmava Berisha A, et al. Vitamin D deficiency 2.0: An update on the current status worldwide. European Journal of Clinical Nutrition. 2020;**74**(11):1498-1513

[4] Griffin MD, Xing N, Kumar R. Vitamin D and its analogs as regulators of immune activation and antigen presentation. Annual Review of Nutrition. 2003;**23**:117-145

[5] Langub MC, Herman JP, Malluche HH, Koszewski NJ. Evidence of functional vitamin D receptors in rat hippocampus. Neuroscience. 2001;**104**(1):49-56

[6] Hovsepian S, Amini M, Aminorroaya A, Amini P, Iraj B. Prevalence of Vitamin D deficiency among adult population of Isfahan City. Iranian Journal of Health Population Nutrition. Apr 2011;**29**(2):149-155

[7] Palacios C, Gonzalez L. Is vitamin D deficiency a major global public health problem? Journal of Steroid Biochemical and Molecular Biology. 2014;**144PA**:138-145

[8] Andıran N, Çelik N, Akça H, Doğan G. Vitamin D deficiency in children and adolescents. Journal of Clinical Research in Pediatric Endocrinology. 2012;**4**(1):25-29

[9] Tabrizi R, Moosazadeh M, Akbari M, Dabbaghmanesh MH, Mohamadkhani M, Asemi Z, et al. High prevalence of Vitamin D deficiency among Iranian ppulation: A systematic review and meta-analysis. Iranian Journal of Medical Science. 2018;**43**(2):125-139

[10] Cashman KD, Dowling KG, Škrabáková Z, Gonzalez-Gross M, Valtueña J, De Henauw S, et al. Vitamin D deficiency in Europe: Pandemic?12. The American Journal of Clinical Nutrition. Apr 2016;**103**(4):1033-1044

[11] Parva NR, Tadepalli S, Singh P, Qian A, Joshi R, Kandala H, et al. Prevalence of Vitamin D Deficiency and Associated Risk Factors in the US Population (2011-2012). Cureus. 2016;**10**(6):e2741

[12] Riley PA. Melanin. The International Journal of Biochemistry & Cell Biology. Nov 1997;**29**(11):1235-1239

[13] Bikle DD, Vitamin D. Metabolism and function in the skin. Molecular and Cellular Endocrinology. 2011;**347**(1-2):80-89

[14] Glatt DU, McSorley E, Pourshahidi LK, Revuelta Iniesta R, McCluskey J, Beggan L, et al. Vitamin D status and health outcomes in school children in Northern Ireland: Year One Results from the D-VinCHI Study. Nutrients. 2022;**14**(4):804

[15] Peterson CA, Belenchia AM. Vitamin D deficiency & childhood obesity: A tale of two epidemics. Missouri Medicine. 2014;**111**(1):49-53

[16] Margulies SL, Kurian D, Elliott MS, Han Z. Vitamin D deficiency in patients with intestinal malabsorption

syndromes – think in and outside the gut: Malabsorption and vitamin D deficiency. Journal of Digestive Diseases. 2015;**16**(11):617-633

[17] Parizadeh SM, Rezayi M, Jafarzadeh-Esfehani R, Avan A, Ghazizadeh H, Emadzadeh M, et al. Association of vitamin D status with liver and kidney disease: A systematic review of clinical trials, and cross-sectional and cohort studies. International Journal for Vitamin and Nutrition Research. 2021;**91**(1-2):175-187

[18] Gröber U, Kisters K. Influence of drugs on vitamin D and calcium metabolism. Dermato-Endocrinology. 2012;**4**(2):158-166

[19] Choudhury S, Kf J, Ranabir S, Singh TP. Familial vitamin D resistant rickets: End-organ resistance to 1,25-dihydroxyvitamin D. Indian Journal of Endocrine Metabolism. 2013;**17**(7):224

[20] Jones G, Prosser DE, Kaufmann M. Cytochrome P450-mediated metabolism of vitamin D. Journal of Lipid Research. 2014;**55**(1):13-31

[21] Zmijewski MA. Vitamin D and human health. International Journal of Molecular Sciences. 2019;**20**(1):145

[22] Bikle DD. Vitamin D: Production, metabolism and mechanisms of action. In: Feingold KR, Anawalt B, Boyce A, Chrousos G, de Herder WW, Dhatariya K, et al., Endotext. South Dartmouth (MA): MDText.com, Inc.; 2000

[23] Nair R, Maseeh A. Vitamin D: The "sunshine" vitamin. Journal of Pharmacology and Pharmacotherapeutics. 2012;**3**(2):118-126

[24] Pilz S, Zittermann A, Trummer C, Theiler-Schwetz V, Lerchbaum E, Keppel MH, et al. Vitamin D testing and treatment: A narrative review of current evidence. Endocrine Connections. 2019;**8**(2):R27-R43

[25] Hany M, Rehman B, Azhar Y, Chapman J. Schizophrenia. Treasure Island (FL): StatPearls Publishing; 2022

[26] Patel KR, Cherian J, Gohil K, Atkinson D. Schizophrenia: Overview and treatment options. P T. Sep 2014;**39**(9):638-645

[27] Brisch R, Saniotis A, Wolf R, Bielau H, Bernstein HG, Steiner J, et al. The role of dopamine in Schizophrenia from a Neurobiological and Evolutionary Perspective: Old Fashioned, but Still in Vogue. Frontiers in Psychiatry. 2014;**5**:47

[28] Mouse Models of Schizophrenia: Risk Genes – ScienceDirect. [cited June 15, 2022]. Available from: https://www.sciencedirect.com/science/article/pii/B978012800981900016X

[29] Olijslagers J, Werkman T, McCreary A, Kruse C, Wadman W. Modulation of midbrain dopamine neurotransmission by serotonin, a versatile interaction between neurotransmitters and significance for antipsychotic drug action. Current Neuropharmacology. 2006;**4**(1):59-68

[30] Chokhawala K, Stevens L. Antipsychotic Medications. Treasure Island (FL): StatPearls Publishing; 2022

[31] Viani-Walsh D, Kennedy-Williams S, Taylor D, Gaughran F, Lally J. Vitamin D deficiency in schizophrenia implications for COVID-19 infection. Iranian Journal of Psychological Medicine:1-10

[32] Yüksel RN, Altunsoy N, Tikir B, Cingi Külük M, Unal K, Goka S, et al. Correlation between total vitamin D levels and psychotic psychopathology in

patients with schizophrenia: Therapeutic implications for add-on vitamin D augmentation. Therapeutic Advances in Psychopharmacology. 2014;**4**(6):268-275

[33] Clelland JD, Read LL, Drouet V, Kaon A, Kelly A, Duff KE, et al. Vitamin D insufficiency and schizophrenia risk: evaluation of hyperprolinemia as a mediator of association. Schizophrenia Research. 2014;**156**(1):15-22

[34] Cecchini NM, Monteoliva MI, Alvarez ME. Proline dehydrogenase is a positive regulator of cell death in different kingdoms. Plant Signaling & Behavior. 2011;**6**(8):1195-1197

[35] Eyles DW. Vitamin D: Brain and behavior. JBMR Plus. 2020;**5**(1):e10419

[36] Cui X, McGrath JJ, Burne THJ, Eyles DW. Vitamin D and schizophrenia: 20 years on. Molecular Psychiatry. 2021;**26**(7):2708-2720

[37] Hawes JE, Tesic D, Whitehouse AJ, Zosky GR, Smith JT, Wyrwoll CS. Maternal vitamin D deficiency alters fetal brain development in the BALB/c mouse. Behavioural Brain Research. 2015;**286**:192-200

[38] O'Loan J, Eyles DW, Kesby J, Ko P, McGrath JJ, Burne THJ. Vitamin D deficiency during various stages of pregnancy in the rat; its impact on development and behaviour in adult offspring. Psychoneuroendocrinology. 2007;**32**(3):227-234

[39] Krivoy A, Onn R, Vilner Y, Hochman E, Weizman S, Paz A, et al. Vitamin D supplementation in chronic Schizophrenia patients treated with clozapine: A Randomized, Double-Blind. Placebo-controlled Clinical Trial. EBioMedicine. 2017;**26**:138-145

[40] Neriman A, Hakan Y, Ozge U. The psychotropic effect of vitamin D supplementation on schizophrenia symptoms. BMC Psychiatry. 2021;**21**(1):309

[41] Gibson PG, Qin L, Puah SH. COVID-19 acute respiratory distress syndrome (ARDS): clinical features and differences from typical pre-COVID-19 ARDS. Medical Journal of Australia. 2020;**213**(2):54-56

[42] Li X, Ma X. Acute respiratory failure in COVID-19: is it "typical" ARDS? Critical Care. 6 May 2020;**24**:198

[43] Ghelani D, Alesi S, Mousa A. Vitamin D and COVID-19: An overview of recent evidence. International Journal of Molecular Sciences. 2021;**22**(19):10559

[44] Pike JW, Meyer MB. The Vitamin D receptor: New paradigms for the regulation of gene expression by 1,25-dihydroxyvitamin D3. Endocrinology and Metabolism Clinics of North America. 2010;**39**(2):255-269

[45] Shirvani A, Kalajian TA, Song A, Holick MF. Disassociation of Vitamin D's calcemic activity and non-calcemic genomic activity and individual responsiveness: A Randomized Controlled Double-Blind Clinical Trial. Scientific Reports. 2019;**9**:17685

[46] Aranow C. Vitamin D and the immune system. Journal of Investigative Medicine. 2011;**59**(6):881-886

[47] Yisak H, Ewunetei A, Kefale B, Mamuye M, Teshome F, Ambaw B, et al. Effects of Vitamin D on COVID-19 infection and prognosis: A systematic review. Risk Management and Healthcare Policy. 2021;**14**:31-38

[48] Gerke AK, Pezzulo AA, Tang F, Cavanaugh JE, Bair TB, Phillips E, et al. Effects of vitamin D supplementation on alveolar macrophage gene expression:

preliminary results of a randomized, controlled trial. Multidisciplinary Respiratory Medicine. 2014;**9**(1):18

[49] Kota SK, Kota SK, Jammula S, Meher LK, Panda S, Tripathy PR, et al. Renin–angiotensin system activity in vitamin D deficient, obese individuals with hypertension: An urban Indian study. Indian Journal of Endocrinology Metabolism. 2011;**15**(Suppl. 4): S395-S401

[50] Ajabshir S, Asif A, Nayer A. The effects of vitamin D on the renin-angiotensin system. Journal of Nephropathology. 2014;**3**(2):41-43

[51] McMullan CJ, Borgi L, Curhan GC, Fisher N, Forman JP. The effect of Vitamin D on renin-angiotensin-system activation and blood pressure – A Randomized Control Trial. Journal of Hypertension. 2017;**35**(4):822-829

[52] Li YC. Vitamin D regulation of the renin-angiotensin system. Journal of Cellular Biochemistry. 2003;**88**(2):327-331

Section 4

Vitamin D Status and Non-Hormonal/Hormonal Disorders

Chapter 8

Hypovitaminosis D in Postmenopause

Patricia Loranca-Moreno, Alan Rios-Espinosa and Juan Moises Ocampo-Godínez

Abstract

Hypovitaminosis D is a common health problem in postmenopausal women that predisposes to the development of various conditions, such as difficult-to-manage osteoporosis, cardiovascular diseases, metabolic syndrome, autoimmune diseases and cancer. In the last two decades, the extensive role of vitamin D has been characterized, where besides controlling bone mineral metabolism, it also precisely regulates the immune system and metabolism. Early detection of hypovitaminosis D can help provide timely care to improve the health of postmenopausal women. This chapter aims to discuss the most relevant aspects of vitamin D in postmenopausal women and the probable consequences that it has on the development of pathological processes characteristic of this stage.

Keywords: postmenopause, hypovitaminosis D, vitamin D sufficiency, vitamin D insufficiency, vitamin D deficiency, post-menopausal osteoporosis, vitamin D immunomodulation

1. Introduction

The reproductive system in women experiences a senescence process called menopausal transition. It leads to an imbalance in estrogenic hormonal regulation that is evident around the fifth to sixth decade of life [1–3]. Ovarian insufficiency during menopausal transition, triggers a series of adaptive physiological responses to estrogen reduction, producing clinical signs and symptoms related to central nervous system disorders, cardio-metabolic changes, musculoskeletal disorders, urogenital and skin atrophy, and sexual dysfunction [4]. After this period, menopause occurs. It is the last menstruation and represents the end of a woman's reproductive life. Nowadays, a demographic aging phenomenon is occurring worldwide, and it has been accompanied by a gender transition [5]. Indeed, the life expectancy of women will reach 82 years in 2025, which implies that one-third of women will have postmenopause [6].

Hypovitaminosis D is a public health problem considered a pandemic. In Mexico, we have a prevalence of hypovitaminosis of 89.79% in postmenopausal women. Hypovitaminosis D status in this population has been associated with musculoskeletal problems and a wide range of acute and chronic diseases, including heart disease, autoimmune diseases, metabolic syndrome, and obesity [7, 8]. Furthermore, it can cause secondary hyperparathyroidism in most patients, increasing bone loss and the risk of developing osteomalacia, and predisposing to fragility-fractures in the elderly [9, 10]. It is thought that vitamin D deficiency, defined as serum concentrations of 25-hydroxy-vitamin D (25OHD) <30 nmol/L or <20 ng/mL, should be corrected. However, we have observed that also vitamin D insufficiency, defined as serum levels <50 nmol/L or <29 ng/mL, must also be corrected. Most international guidelines define vitamin D sufficiency as serum levels of 25OHD >50 nmol/L or >29 ng/mL to achieve optimal bone health in older adults. However, we have found that sufficiency levels should even be considered between 40-50 ng/mL of 25 OHD. In selected populations, randomized controlled trials (RCTs) with vitamin D and calcium supplementation, have shown a decrease in the incidence of hip fractures and non-vertebral fractures by ~15%, with the greatest effect in people 80 years or older [11, 12].

2. Vitamin D as a hormone

Formally, vitamin D is defined as a non-essential nutrient. However, It should be considered a hormone because of its biotransformation and transport, and the multiple functions it has in various enzymatic, metabolic, physiological, and even pathophysiological processes [13]. This concept makes the skin and liver the primary glands for the synthesis of vitamin D. As a result, cholecalciferol should be considered a prohormone, transformed into its active form by those organs that express the CYP27B1 enzyme, (mainly the kidney and the immune system), changing Cholecalciferol to Calcitriol. Considering the above, and adding the mechanism of action of vitamin D, this system would be quite reminiscent of the axis of thyroid hormones. Therefore, the concept of "Hormone D" should be considered a serious proposal [14, 15].

2.1 Rich sources of vitamin D

As vitamin D is considered a non-essential nutrient instead of a hormone, it is thought to mostly be acquired through diet. However, it is mainly obtained through an endogenous synthesis that begins in the skin after sunlight exposure. Nonetheless, it is important to address the main nutritional sources that provide vitamin D, since they provide the remaining 10% of daily requirements. The addition of vitamin D in dairy products was effective in the 1930s in the attempt to eradicate Rickets. Adolf Windaus, the Nobel Prize in Chemistry in 1928, had already described this process, in the studies on the constitution of sterols and their connection with vitamins [16, 17]. As we mentioned above, the major source of vitamin D is the production in the skin, which synthesizes 90% of the total vitamin D that we require daily. The rest is obtained through the diet. Plant-based products mainly provide ergocalciferol or vitamin D2, while animal-based products provide vitamin D3. Although current diets are theoretically considered sufficient in vitamin D, many countries have a high prevalence of malnutrition, like Mexico and other Latin American countries; so, this assumption cannot be generalized. As we have already mentioned, the diet only provides a small proportion of the daily needs of vitamin D. If we depended only on the diet completely to accomplish the daily vitamin D requirements, we would face serious problems since the amount of vitamin D in food is very low. For instance, the amount of vitamin D in international units (IU) in egg yolk is 44 IU, if the total

Food and portion	IU of Vitamin D	Amount per day to obtain 4000 IU
Cod liver oil, 1 tablespoon	1360 IU	4 spoons
Cooked trout, 3 ounces	645 IU	18 ounces
Cooked salmon, 3 ounces	570 IU	21 ounces
Mushrooms, white, raw, sliced, exposed to UV light, ½ cups	366 IU	5 ½ cups
Milk, 2% milkfat, fortified with vitamin D,1 cup	120 IU	33 cups
Sardines (Atlantic), canned in oil, drained, 2 sardines	46 IU	172 sardines
1 egg yolk	44 IU	90 eggs
Tuna, canned in water, 3 ounces	40 IU	300 ounces
Cheddar cheese, 1.5 ounces	12 IU	500 ounces

Adaptation from: [18].

Table 1.
Food requirements to meet the daily amount of vitamin D.

needs depended on the diet, it would require 90 eggs per day to get almost 4000 IU of vitamin D from the diet, something unreasonable. **Table 1** shows some examples of these relationships [19–22].

The relevance of sun exposure comes from the impossibility of achieving sufficient vitamin D requirements solely through diet. However, the question would be, how much sun exposure is required to synthesize optimal vitamin D levels per day. That is known as the minimum erythema dose (MED), defined as the dose of ultraviolet B radiation (UV-B) radiation necessary to produce perceptible erythema after 24 hours of irradiation in an exposed subject. Exposure equivalent to 1/4 of the MED on the face, hands and arms produces approximately 1000 IU of vitamin D, while exposure equivalent to 1/6 produces 200 to 600 IU [23]. We have observed that at least 4000 IU daily of vitamin D are needed to reach sufficient serum levels of 25(OH) between 30-50 ng/mL. However, as we can analyze, we need a rich diet in vitamin D and high sun exposure, difficult to achieve because of several social issues. Moreover, high exposure to the sun's rays could also predispose to developing melanoma. Therefore, the daily use of vitamin D supplements could be the best source to get the daily requirements.

2.2 Vitamin D synthesis

Being endogenous synthesis the most important way to obtain vitamin D, we are going to explain the process briefly. After exposure to UV-B rays with a wavelength of 290 to 320 nm in the skin (basal layer of the epidermis), pre-vitamin D3 is synthesized from 7-dehydrocholesterol or pro-vitamin D, which is thermally unstable; therefore, it is spontaneously isomerized into vitamin D3. On the other hand, vitamin D2 and D3 obtained from the diet, are absorbed in the small intestine with the help of bile acids and later transported to the liver. The blood transport of vitamin D2 and vitamin D3 is done through the vitamin D-binding protein (DBP). Once in the liver, the 25 (OH) or vitamin D3 is bio-transformed through enzymes from the P450 complex that includes the CYP2R1, CYP3A4, and CYP2J3 [24, 25].

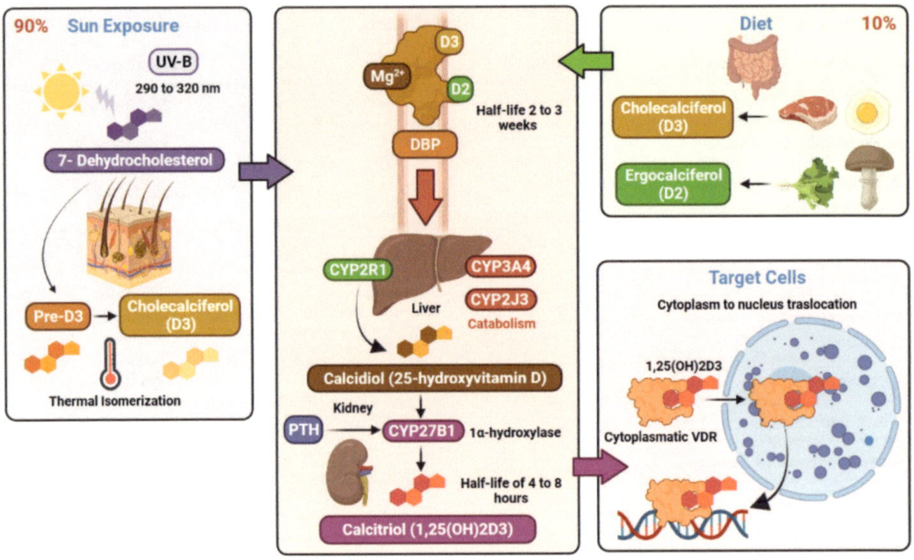

Figure 1.
Vitamin D synthesis pathway.

The main circulating form of vitamin D is vitamin D3, which has a half-life of approximately 2-3 weeks in the blood. This inactive form is transported from the liver to the kidney by DBP. Then, it is transformed into the active form in the proximal tube, the 1-alfa,25-dihidroxicolecalciferol (1,25 (OH) 2 D3) or calcitriol that has a half-life of 4 to 8 hours. This biotransformation is carried out by 1α-hydroxylase, also known as CYP27B1. This is expressed in renal tubular cells after stimulation by the parathormone (PTH). CYP27B1 synthesis is inhibited by high levels of calcium (Ca), phosphorous (P), fibroblast growth factor 23 (FGF23), and 1,25 (OH) 2 D3 itself by the action of 24-hydroxylase that produces 24,25 (OH) 2 vitamin D3 [26–28]. The 1,25 (OH) 2 D3 finally interacts with the vitamin D receptor (VDR) to exert its physiological function. Magnesium (Mg^{2+}) plays an essential role in some of the previously identified steps, such as the binding of vitamin D2, D3, and 1,25 (OH) 2 D3 to DBP, as well as the renal and hepatic hydroxylation for producing 25 (OH) D and the 1,25 (OH) 2 D. Thus, a magnesium deficiency could be relevant for the synthesis and transportation of vitamin D (**Figure 1**) [29].

2.3 Mechanism of action

After obtaining the bio-active form of vitamin D, the interaction with the vitamin D receptor (VDRs) is essential to carry out all its functions. VDRs were initially described in the cytoplasm and the nucleus. Still, they were also found in some fundamental organelles, such as mitochondria [30]. This location reinforced the notion of non-genomic effects on vitamin D. Firstly, 1,25 (OH) 2 D3 interacts with the VDR located in the cytoplasm to form a complex VD/VDR. Then, VD/VDR complex is translocated to the cell nucleus to produce the modulation of multiple target genes, especially those that mediate the bone-mineral metabolism [31]. However, the 1,25 (OH) 2 D3 also regulates genes associated with malignant cell potency, hormone secretion, cytokines, and transcription factors that modulate immune cells. It is important to emphasize that estrogens increase VDRs in certain cells, such as osteoblasts, osteoclasts, and immune cells [9]. Therefore, there is a close relationship between estrogen regulation and vitamin D responsiveness, which could have important consequences in conditions of estrogen deficiency, such as postmenopause.

The VDR gene has many copies in each subject, with several polymorphisms. The main polymorphism modifications occur in promoter and exon regions that have been studied using restriction enzymes, such as BsmI, ApaI, TaqI, and FokI [32]. People could be homozygous and heterozygous for any polymorphism. Indeed, each person has a combination of them due to their co-dominance property, similar to what happens with the Major Histocompatibility Complex (MHC). Hence, haplotype is the name given to the combination of the different polymorphisms in each person. Each different haplotype determines the responsiveness of the cells to vitamin D and can predispose them to develop some diseases, such as cancer or autoimmune diseases [33–37]. The recessive VDR-Bmsl GA/GG polymorphism and the heterozygous recessive VDR-Fokl Ff/FF polymorphism predispose to developing breast cancer. The homozygous dominant VDR-Bmsl BB polymorphism, the homozygous dominant VDR-Fokl FF polymorphism, and the homozygous dominant VDR-CDX2 GG polymorphism predispose to developing ovarian cancer [38]. Also, the fF/bB+ff/BB+FF/bb haplotype is more frequent in women with gestational hypertension and vitamin D deficiency [39]. Regarding the non-genomic effect, it is an action that takes place shortly, such as the mobilization of calcium contained in intracellular vesicles, or the activation of enzymes that metabolize phosphatidylinositol acid.

Moreover, recent studies have shown non-classical extraosseous effects of vitamin D, which are mainly related to immunomodulation and could play an important role in different autoimmune and autoinflammatory diseases, such as Type 1 Diabetes Mellitus, Systemic Lupus Erythematosus (SLE), Rheumatoid Arthritis (RA), Inflammatory Bowel Disease (IBD), Psoriasis, among others [40, 41].

3. Postmenopause and hypovitaminosis D

As vitamin D is fat-soluble and requires bile salts to be absorbed in the duodenum, malabsorption syndromes could have significant consequences in this process. Therefore, the absorption of vitamin D is significantly affected by short bowel syndrome, celiac disease, cystic fibrosis, pancreatic insufficiency, and cholestatic liver diseases. Likewise, it is affected by the low intake of vitamin D in the elderly population and the amount of vitamin D in the different types of diets [42, 43].

As a result of current human behavior (long stays in hospital, the use of sunscreen, religious reasons), there is a significant deficiency of vitamin D. The type of skin also decreases the synthesis of vitamin D since melanin is a natural blocker of ultraviolet rays, slowing down the production of vitamin D. Also, age decreases vitamin D synthesis in the skin, whereby hypovitaminosis is very marked in postmenopausal women [44–46].

Chronic liver diseases such as cirrhosis and liver failure may have a defective hydroxylation that leads to a lack of vitamin D activation.

Likewise, renal biotransformation can be affected by hyperparathyroidism, renal failure, 1-α hydroxylase deficiency, and in elderly patients whose hydroxylation is decreased due to an idiopathic cause. Some drugs can induce hepatic p450 enzymes, which leads to increased vitamin D degradation, including phenobarbital, carbamazepine, tamoxifen, rifampin, spironolactone, dexamethasone, nifedipine, and clotrimazole, which activate the Pregnane X receptor (PXR) [47].

Obese people have a higher risk of vitamin D deficiency due to the decrease in its availability; since vitamin D is a fat-soluble vitamin, it might be sequestered in body fat. Moreover, adipose tissue has VDR receptors, so it could act as a sponge that also sequesters 80% of serum vitamin D, especially when the waist perimeter exceeds 85 cm [48, 49].

Moreover, the production of vitamin D could be affected by some polymorphisms, mutations, or epigenetic changes in genes associated with endogenous production, such as the gene of the CYP27B1 enzyme.

3.1 Postmenopausal osteoporosis and hypovitaminosis D

Osteoporosis is a condition most favored by the state of hypovitaminosis D in postmenopause. According to the National Institutes of Health Consensus Development Panel on Osteoporosis, it is defined as "a skeletal disorder characterized by compromised bone strength leading to an increased risk of fracture." According to the criteria of the World Health Organization (WHO), osteoporosis is defined as a bone mineral density that is 2.5 or more standard deviations (SD) below average [50, 51]. In February 2010, the United States Preventive Services Task Force conducted a systematic review of trials published and concluded that vitamin D reduced falls by 17% (95% CI, 11%-23%) over 6 to 36-month follow-up [52]. In the last decade, commercial assays for 25(OH)D have become widely available, allowing researchers to easily measure vitamin D stores in people. Subsequently, many research studies indicated that higher levels of 25(OH)D were associated with higher calcium absorption efficiency, lower risk of secondary hyperparathyroidism, higher bone mineral density, and lower risk of fractures. Physiopathological situations can be associated with the natural effect of aging and secondary hypoestrogenism (**Figure 2**).

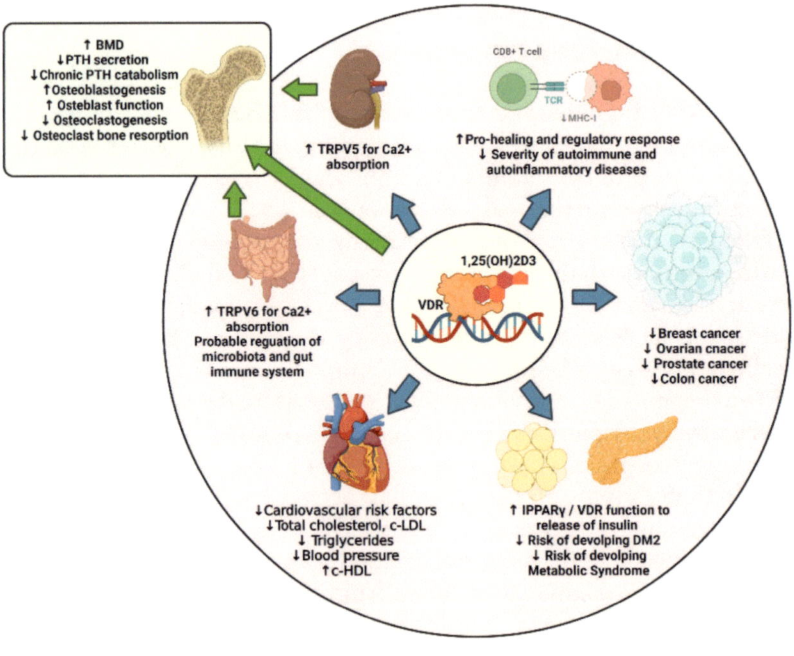

Figure 2.
Effects of vitamin D on different organs and tissues.

In 2005, six experts reviewed the literature and published a paper indicating that the optimal vitamin D level for bone health was 30 ng/mL or greater (75 nmol/L). However, others stated that a 25(OH)D level of 20 ng/mL or greater (50 nmol/L) was adequate. Laboratories across the country adopted 30 ng/mL as the new threshold showing regular stores of vitamin D.

The Institute of Medicine will improve a vitamin D intake of 400 IU/d for people between the ages of 0 and 1 year, 600 IU/day for people between 1 and 70 years (including pregnant and lactating women), and 800 IU/day for those older than 70 years [50]. However, we have already mentioned that most of the beneficial effects of vitamin D, both in bone-mineral metabolism and in the immune system, are reached when serum levels get between 30-50 ng/ml.

3.2 Vitamin D supplementation remit hyperparathyroidism

Intermittent pulses of parathyroid hormone (PTH) have shown to be osteo-anabolic. However, chronic exposure to high levels of PTH, like hyperparathyroidism, is catabolic for bone. Vitamin D therapy can reduce PTH levels and improve bone mineral density (BMD). Therefore, the serum level of 25(OH)D that minimizes PTH is relevant for bone health, particularly among individuals with normal renal function. Some authors concluded that the risk of secondary hyperparathyroidism was significantly reduced when serum 25(OH)D levels exceeded approximately 20-30 ng/ml [50].

Our experience has demonstrated that most of our post-menopausal patients with hypovitaminosis D also suffer from hyperparathyroidism and all respond to vitamin D supplementation. Indeed, we have had patients with levels above 30 ng/ml with hyperparathyroidism that also remit after vitamin D supplementation. Interestingly, the remission of hyperparathyroidism avoids the catabolism of bone and improves BMD in patients with severe osteoporosis [53]. Thereby, this is why we have questioned the current cut-off levels of "vitamin D sufficiency". Therefore, vitamin D sufficiency should be considered only when vitamin D serum levels are between 40-50 ng/ml, especially in postmenopausal women.

Currently, it exists a controversy regarding the dose of complementary administration of vitamin D. The Osteoporosis Clinical Practice Guidelines in Mexico recommend the administration of 1000 mg of calcium and 800 IU of vitamin D per day in women with postmenopausal osteoporosis. However, we have observed that supplementation with 4000 IU daily achieves an improvement in serum levels of vitamin D (41.7 ng/dL) with no adverse effects.

3.3 Glucose, vitamin D and Type 2 Diabetes Mellitus (DM2)

In obese postmenopausal women, lipid peroxidation can promote the sequestration of vitamin D in adipose tissue, increasing the risk of developing DM2 [54]. There are several mechanisms that could explain the association between glucose alterations, diabetes mellitus, and vitamin D, such as the relationship between the VDR and 1-α-hydroxylase in beta cells of the pancreas, as well as the interaction between the peroxisome proliferator-activated receptor γ (PPARγ) and the VDR to release insulin [55].

3.4 Biochemical and clinical markers of cardiovascular risk related to hypovitaminosis D

Dyslipidemia and high blood pressure are important cardiovascular risk factors in postmenopausal women. A recently published meta-analysis including 81 studies, suggests that vitamin D supplementation is useful in protecting against cardiovascular disease by improving risk factors, including high blood pressure, hyperparathyroidism, dyslipidemia, and chronic inflammation. There were observed beneficial effects on lipid levels and high-sensitivity C-reactive protein (hs-CRP) after vitamin D supplementation. A dose of 3000 IU/day of vitamin D, showed a significant reduction in total cholesterol, LDL cholesterol (cLDL), and triglycerides, with an increase in HDL cholesterol (cHDL). Subgroup analysis showed that the effect on triglycerides and cHDL was more significant in participants who received vitamin D supplementation for ≥6 months. hs-CRP concentrations were slightly lower with vitamin D doses ≥4000 IU/day compared with lower doses [56].

Arterial hypertension is the main cardiovascular risk factor that affects women. Cardiovascular aging is accelerated by the presence of risk factors that appear in postmenopause. The increased stiffness of the large arteries results in a larger pulse wave that increases arterial hypertension [57].

Some studies suggest that vitamin D deficiency can predispose to developing high blood pressure. Possible mechanisms in the association of vitamin D and blood pressure include an inverse relationship between vitamin D concentrations and the renin-angiotensin-aldosterone system (RAAS), as well as the prevention of secondary hyperparathyroidism [58]. Furthermore, high levels of PTH are related to vitamin D deficiency, which can be related to myocardial hypertrophy and elevated blood pressure.

4. Vitamin D and the immune system

Vitamin D plays an important role in the function of the immune system. The cessation of estrogens is typical in menopause and affects the function of the immune system, favoring a change in the basal immune response. After this event, the systemic immune response tends to be polarized towards a mixed type 1 and type 3 pro-inflammatory response, which is associated with the pathogenesis of multiple diseases. Type 1 and type 3 responses could be essential in the pathophysiology of various conditions in postmenopause. The mixed pro-inflammatory response that we previously mentioned generates an increase in other pro-inflammatory serum cytokines, such as TGF-β, adiponectin, adipsin, plasminogen activator inhibitor-1 (PAI), IL-6, IL-7, IL-12, IL-17 and IL-23, and decreases cytokines associated with a regulatory response or pro-healing type 2 response, such as IL-4 and IL-10 (**Figure 3**) [59–70].

Regarding the adaptive immune system cells, the number of CD4 T cells decreases with a predominance of T helper cells (Th) with Th1 and Th17 phenotypes. In addition, the number of CD8 T cells increases, causing a reversal of the CD4/CD8 cell ratio [71]. It is still to be discerned the importance of the type 3 response in the pathophysiology of classic postmenopausal disorders. Nevertheless, the involvement of this mixed pro-inflammatory response could explain many cases of treatment-resistant osteoporosis that have occurred in our clinic. The treatment of osteoporosis is often focused on inhibiting the action of the osteoclast and the RANKL pathway to enhance the synthesis of bone mineral matrix by osteoblasts. However, the osteoclastic and osteoblastic functions can also be regulated by other cytokines and cells from both type 1 and type 3 responses (**Figure 4**). Studies in murine models with monoclonal antibodies anti-IL-17 and anti-IL-23 have demonstrated to be capable of reducing the expression of pro-inflammatory cytokines type 1 and 3 and increasing the BMD, either

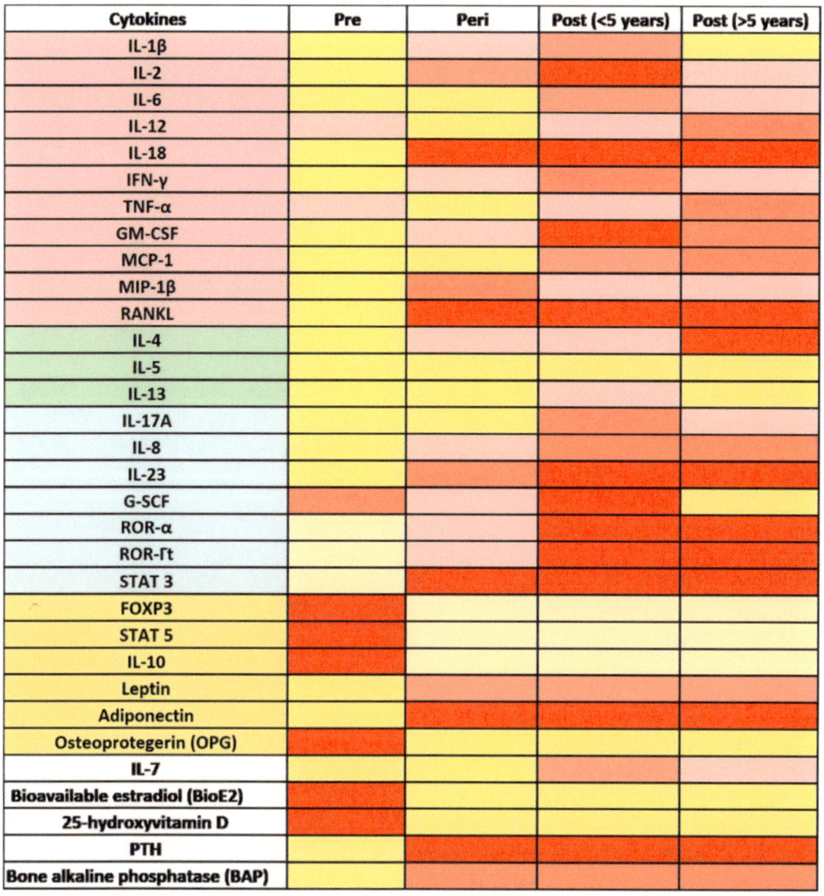

Figure 3.
Heat map of cytokine impairment during perimenopause and post-menopause.

of cancellous and cortical bone [72]. Anti-IL-23 treatment also inhibits osteoclastogenesis and promotes osteogenic differentiation of mesenchymal stem cells (MSCs).

Also, anti-IL-17 therapy has shown a greater improvement in BMD and a greater decrease in type 1 and type 3 cytokines, compared to monoclonal anti-RANKL and anti-TNFα [73]. Although hormone therapy has shown an effect in reducing IL-2, TNF-α, and even IL-17, it does not completely reverse the effects of immune dysregulation.

4.1 The non-classical role of vitamin D on the immune system and the consequences of hypovitaminosis in postmenopausal women

This impairment in the immune response has also been associated with a state of hypovitaminosis D and hyperparathyroidism in post-menopause [74–76]. Vitamin D is an essential hormone to maintain the homeostasis of bone mineral metabolism. However, vitamin D also plays a vital role in maintaining a proper immune response. The active form of calcitriol has autocrine and paracrine effects on both innate and adaptive immune cells.

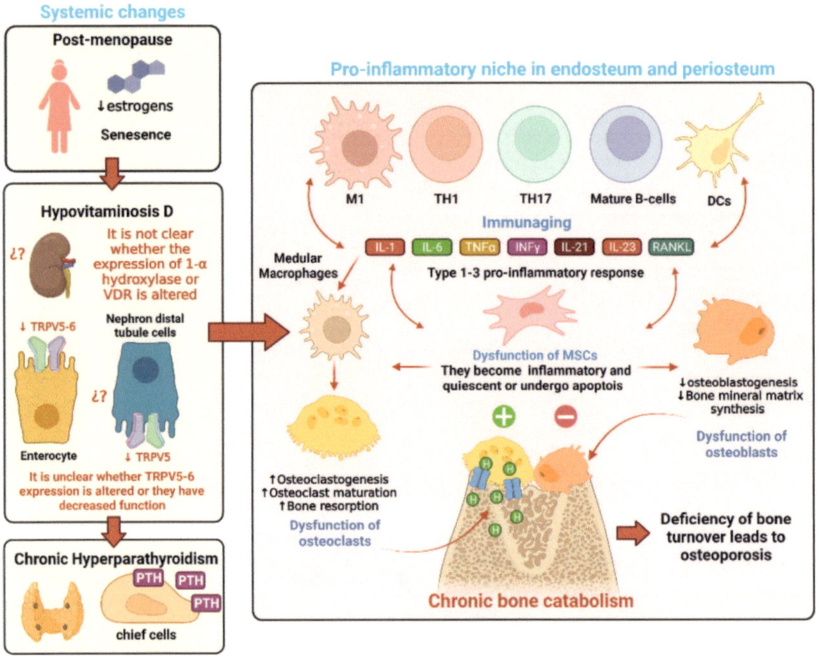

Figure 4.
The effect of the mixed pro-inflammatory response type 1 and type 3 in bone homeostasis.

Antigen-presenting cells (APCs), such as macrophages, dendritic cells, and B cells, express the enzyme 1-α-hydroxylase, also known as CYP27B1, an enzyme previously thought to be expressed only by renal tubular cells. This enzyme converts 25 (OH) to 1,25 (OH)2 D3 or calcitriol, the active form of vitamin D [77]. Contrary, T cells express a very low amount of this enzyme, thus requiring APCs to bio-transform 25 (OH) into 1,25 (OH)2 D3, illustrating the important role of APCs in regulating the bioavailability of calcitriol for T cells. Interestingly, CYP27B1 expression on APCs is increased after contact with their major histocompatibility complexes (MHCs) with T cell receptors (TCRs). This means that APCs can always use vitamin D, but in the case of T cells, vitamin D metabolism and utilization, only efficiently occur after antigen presentation (**Figure 5**) [78].

To understand better the function of vitamin D in the immune system, it is essential to remember the interaction between APCs and T cells, which comprises three signals wonderfully explained by Janeway [79]. The first signal corresponds to the initial contact of T-cells with APCs through the TCR and the MHC, besides the interaction of CD4 or CD8 co-receptors. The second signal comprises stimulation through the interaction of CD80 or CD86 with CD28, enhancing the state of lymphocyte activation. These interactions occur a dozen times within a close contact zone between the APC and the lymphocyte, called the immunological synapse. It is well known that these two signals trigger many downfall pathways either in the APCs and T cells. Regarding APCs, these pathways trigger the transcription of many cytokines, constituting the third signal, which helps to polarize the stimulated T-cells towards the distinct Th phenotypes. Moreover, these signals are necessary to enhance the transcription of the CYP27B1 enzyme, which stimulates the production of calcitriol in APCs, being released to lymphocytes with the rest of the cytokines as part of the third signal [80]. In this context, the stimulation of 1,25 (OH)2 D3

DOI: http://dx.doi.org/10.5772/intechopen.106716

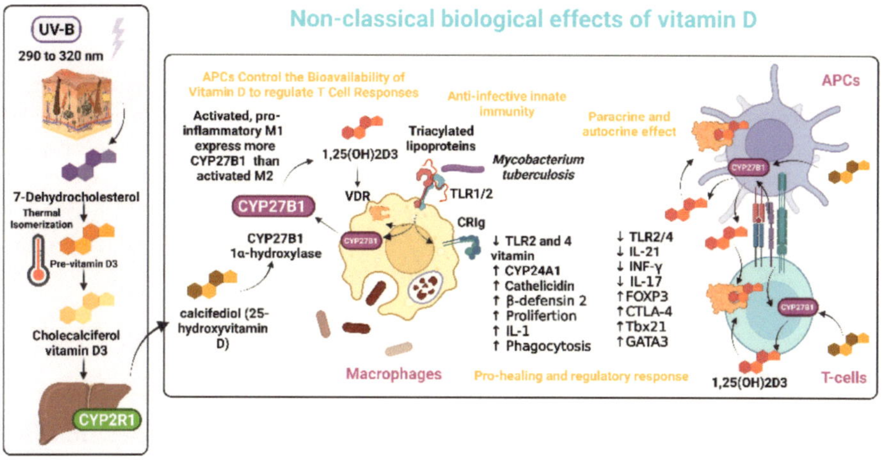

Figure 5.
The role of vitamin D in both innate and adaptative immune systems.

Decreases the production, of type I and type 3 proinflammatory cytokines, such as IL-21, IFN-γ, and IL-17, and stimulates the production of transcription factors and regulatory cytokines (such as FOXP-3, IL-10, and CTLA-4), as well as transcription factors that polarize the response towards a pro-healing type 2 response (such as GATA-3). Likewise, the autocrine stimulation of vitamin D on APCs produces a decrease in Toll-like receptors 2 and 4 and decreases their proliferation. Furthermore, calcitriol increases the production of CYP24A, cathelicidin, β-defensin 2, and IL-1. Together, these phenomena enhance the antimicrobial activity and phagocytosis of APCs without triggering an exaggerated inflammatory response, showing the undoubted capacity of immunomodulation that vitamin D possesses (**Figure 5**) [77–81].

This information suggests that vitamin D plays an essential role in every immune response, works as a handbrake to prevent excess production of proinflammatory cytokines, and helps to produce regulatory cytokines, optimizing the clearance process of pathogens or damaged products and promoting tissue repair and homeostasis. The immunomodulatory effect of vitamin D on type 1 and 3 responses helps to explain why high-dose vitamin D supplementation improves the severity of signs and symptoms in cohorts of patients with SLE and RA. Likewise, it aids in explaining why vitamin D has a protective effect on patients at risk of developing breast and ovarian cancer. In addition, it is possible to understand the relevance of hypovitaminosis D in cohorts of postmenopausal patients by knowing the role of vitamin D in the immune system.

This can have consequences on the bone and also on the immunological system, explaining why numerous women with post-menopausal osteoporosis do not show significant improvement in BMD with antiresorptive therapies or even with the monoclonal therapies based on RANKL neutralization until they receive supplementation with high doses of vitamin D. Moreover, improvement of hard-to-treat osteoporosis with vitamin D supplementation could be due to immunomodulation of osteoclastic function and other immune cells associated with bone marrow, endosteum and periosteum [82–87]. Future studies will be necessary to continue learning more about the relevance of immuno-endocrine effect of vitamin D. However, it is time to break with the dogma and start treating the hypovitaminosis D in our patients to achieve a proper performance of their immune systems.

5. Conclusion

Vitamin D or "Hormone D", has multiple functions on various tissues and organs, especially in the immune system. This important regulation over the immune system could explain why a deficiency of this hormone can predispose to the development of some pathological conditions in postmenopause women, such as osteoporosis, cancer, metabolic, and cardiovascular diseases. It is necessary to continue carrying out more studies to know the physiological scope of vitamin D, as well as to finish understanding the therapeutic benefits of reaching optimal levels of vitamin D.

Acknowledgements

We want to express our thanks to Professor Gabriela Quintana Fonseca for the language revision.

Conflict of interest

The authors declare no conflict of interest.

Note

All the images were created with BioRender.com.

Abbreviations

APCs	Antigen-Presenting Cells
BAP	Bone Alkaline Phosphatase
BioE2	Bioavailable Estradiol
BMD	Bone Mineral Density
Ca^{2+}	Calcium
CD	Cluster of Differentiation
cHDL	HDL cholesterol
cLDL	LDL cholesterol
CRIg	Complement Receptor of Immunoglobulin superfamily
CYP	Cytochrome P450
CTLA	4-Cytotoxic T-Lymphocyte Antigen 4
DBP	Vitamin D Binding Protein
Dc	Dendritic cells
DM2	Type 2 Diabetes Mellitus
D2	Ergocalciferol
D3	Cholecalciferol
FGF23	Fibroblast Growth Factor 23
FOXP3	Forkhead Box P3
GATA3	GATA-binding protein 3 to DNA sequence [A/T]GATA[A/G]
G-CSF	Granulocyte Colony-Stimulating Factor

GM-CSF	Granulocyte-macrophage colony-stimulating factor
hs-CRP	High-sensitivity C-reactive protein
IBD	Inflammatory Bowel Disease
IL	Interleukin
INF-γ	interferon gamma
IU	International Units
MCP-1	Monocyte Chemoattractant Protein-1
MED	Minimum Erythema Dose
Mg^{2+}	Magnesium
MHC	Major Histocompatibility Complex
MHCs	Major Histocompatibility Complexes
MIP	1alpha-Macrophage Inflammatory Protein-1beta
MSCs	Mesenchymal Stem Cells
OH	Hydroxyl group
OPG	Osteoprotegerin
P	Phosphorous
PAI	Plasminogen Activator Inhibitor-1
PPARγ	Proliferator-Activated Receptor γ
PTH	Parathormone
PXR	Pregnane X Receptor
RAAS	Renin-Angiotensin-Aldosterone System
RA	Rheumatoid Arthritis
RANKL	Receptor Activator for Nuclear Factor κ B Ligand
RCTs	Randomized Controlled Trials
ROR α	Retinoid-related orphan receptor alpha
ROR γ	Retinoid-related orphan receptor gamma
SD	Standard Deviation
SLE	Systemic Lupus Erythematosus
STAT	Signal Transducer and Activator of Transcription
TBX21	T-Box Transcription Factor 21
TCR	T cell receptor
TCRs	T cell receptors
TGF-β	Transforming growth factor β
Th	T Helper cells
TLR	Toll-like Receptor
TLRs	Toll-like Receptors
TNF-α	Tumor Necrosis Factor α
TRPV5	Transient receptor potential cation channel subfamily V member 5
TRPV6	Transient receptor potential cation channel subfamily V member 6
UV-B	Ultraviolet-B
VD	Vitamin D
VDR	Vitamin D receptor
WHO	World Health Organization
1,25 (OH)2 D3	1-alfa,25-dihidroxicolecalciferol
25OHD	25-hydroxy-vitamin D

Author details

Patricia Loranca-Moreno[1*], Alan Rios-Espinosa[1] and Juan Moises Ocampo-Godínez[2]

1 Peri-Postmenopause and Bone Metabolism Clinic, Regional Hospital 1st of October—Institute of Security and Social Services for State Workers (ISSSTE). Av. National Polytechnic Institute, Mexico City, Mexico

2 Tissue Bioengineering Laboratory, National Autonomous University of Mexico (UNAM), Mexico City, Mexico

*Address all correspondence to: paty_lm_2502@hotmail.com

References

[1] Blümel JE, Lavín P, Vallejo MS, Sarrá S. Menopause or climacteric, just a semantic discussion or has it clinical implications? Climacteric. 2014;**17**(3):235-241. DOI: 10.3109/13697137.2013.838948

[2] El Khoudary SR, McClure CK, Vopham T, Karvonen-Gutierrez CA, Sternfeld B, Cauley JA, et al. Longitudinal assessment of the menopausal transition, endogenous sex hormones, and perception of physical functioning: The study of women's health across the nation. The Journals of Gerontology. Series A, Biological Sciences and Medical Sciences. 2014;**69**(8):1011-1017. DOI: 10.1093/gerona/glt285

[3] Jones CM, Boelaert K. The endocrinology of ageing: A mini-review. Gerontology. 2015;**61**(4):291-300. DOI: 10.1159/000367692

[4] Wang TJ. Vitamin D and cardiovascular disease. Annual Review of Medicine. 2016;**67**:261-272. DOI: 10.1146/annurev-med-051214-025146

[5] Monteleone P, Mascagni G, Giannini A, Genazzani AR, Simoncini T. Symptoms of menopause—Global prevalence, physiology and implications. Nature Reviews. Endocrinology. 2018;**14**(4):199-215. DOI: 10.1038/nrendo.2017.180

[6] Rosas-Peralta M, Holick MF, Borrayo-Sánchez G, Madrid-Miller A, Ramírez-Árias E, Arizmendi-Uribe E. Dysfunctional immunometabolic effects of vitamin D deficiency, increased cardiometabolic risk. Potential epidemiological alert in America? Endocrinologia, Diabetes y Nutricion. 2017;**64**(3):162-173. DOI: 10.1016/j.endinu.2016.11.009

[7] Takahashi TA, Johnson KM. Menopause. The Medical Clinics of North America. 2015;**99**(3):521-534. DOI: 10.1016/j.mcna.2015.01.006

[8] Sozen T, Ozisik L, Calik BN. An overview and management of osteoporosis. European Journal of Rheumatology. 2017;**4**(1):46-56. DOI: 10.5152/eurjrheum.2016.048

[9] Bouillon R, Marcocci C, Carmeliet G, Bikle D, White JH, Dawson-Hughes B, et al. Skeletal and extraskeletal actions of Vitamin D: Current evidence and outstanding questions. Endocrine Reviews. 2019;**40**(4):1109-1151. DOI: 10.1210/er.2018-00126

[10] Walker MD, Saeed I, Lee JA, Zhang C, Hans D, Lang T, et al. Effect of concomitant vitamin D deficiency or insufficiency on lumbar spine volumetric bone mineral density and trabecular bone score in primary hyperparathyroidism. Osteoporosis International. 2016;**27**(10):3063-3071. DOI: 10.1007/s00198-016-3637-0

[11] Lee JH, Kim JH, Hong AR, Kim SW, Shin CS. Skeletal effects of vitamin D deficiency among patients with primary hyperparathyroidism. Osteoporosis International. 2017;**28**(5):1667-1674. DOI: 10.1007/s00198-017-3918-2

[12] Alonso Lebrero E, Manuel Barat Baviera J, Pilar Conchello Moreno M, Estruch Riba R, An-tonia Ferrús Pérez M, Font Pérez G, et al. Informe del Comité Científico de la Agencia Española de Consumo, Seguridad Alimentaria y Nutrición (AECOSAN) en relación a la com-plementación con vitamina D de la dieta de niños de 0 a 3 años; 18 de Noviembre del 2015. España. 2015;**2015**:133-150

[13] Jorde R. RCTS are the only appropriate way to demonstrate the role of vitamin D in health. The Journal of Steroid Biochemistry and Molecular Biology. 2018;**177**:10-14. DOI: 10.1016/j.jsbmb.2017.05.004

[14] Lips P. Vitamin D physiology. Progress in Biophysics and Molecular Biology. 2006;**92**(1):4-8. DOI: 10.1016/j.pbiomolbio.2006.02.016

[15] Sempos CT, Heijboer AC, Bikle DD, Bollerslev J, Bouillon R, Brannon PM, et al. Vitamin D assays and the definition of hypovitaminosis D: Results from the First International Conference on Controversies in Vitamin D. British Journal of Clinical Pharmacology. 2018;**84**(10):2194-2207. DOI: 10.1111/bcp.13652

[16] Davies P, Brown RC, Woodhead JS. Serum concentrations of vitamin D metabolites in untreated tuberculosis. Thorax. 1985;**40**(3):187-190. DOI: 10.1136/thx.40.3.187

[17] Wolf G. The discovery of Vitamin D: The Contribution of Adolf Windaus. The Journal of Nutrition. 2004;**134**:1299-1302. DOI: 10.1093/jn/134.6.1299

[18] Vitamin D Content of Selected Foods [Internet]. 2022. Available from: https://ods.od.nih.gov/factsheets/VitaminD-HealthProfessional/#en25

[19] Brandi ML. Indications on the use of vitamin D and vitamin D metabolites in clinical phenotypes. Clinical Cases in Mineral and Bone Metabolism. 2010;**7**(3):243-250. PMID: 22460535

[20] Dominguez LJ, Farruggia M, Veronese N, Barbagallo M. Vitamin d sources, metabolism, and deficiency: Available compounds and guidelines for its treatment. Metabolites. 2021;**11**(4):1104

[21] Durrant LR, Bucca G, Hesketh A, Möller-Levet C, Tripkovic L, Wu H, et al. Vitamins D2 and D3 have overlapping but different effects on the human immune system revealed through analysis of the blood transcriptome. Frontiers in Immunology. 2022;**24**(13):790444

[22] Zhang S, Miller DD, Li W. Non-musculoskeletal benefits of Vitamin D beyond the musculoskeletal system. International Journal of Molecular Sciences. 2021;**22**(4):2128. DOI: 10.3390/ijms22042128

[23] Neville JJ, Palmieri T, Young AR. Physical Determinants of Vitamin D Photosynthesis: A Review. Blackwell Publishing Ltd; 2021

[24] Christakos S, Dhawan P, Verstuyf A, Verlinden L, Carmeliet G. Vitamin D: Metabolism, molecular mechanism of action, and pleiotropic effects. Physiological Reviews. 2016;**96**(1):365-408. DOI: 10.1152/physrev.00014.2015

[25] Baldock PA, Thomas GP, Hodge JM, Baker SUK, Dressel U, O'Loughlin PD, et al. Vitamin D action and regulation of bone remodeling: Suppression of osteoclastogenesis by the mature osteoblast. Journal of Bone and Mineral Research. 2006;**21**(10):1618-1626. DOI: 10.1359/jbmr.060714

[26] Valero Zanuy MÁ, Hawkins CF. Metabolismo, fuentes endógenas y exógenas de vitamina D. Revista Española de Enfermedades Metabolicas Oseas. 2007;**16**(4):63-70. DOI: 10.1016/S1132-8460(07)73506-7

[27] van Driel M, Koedam M, Buurman CJ, Roelse M, Weyts F, Chiba H, et al. Evidence that both 1α,25-dihydroxyvitaminD3and24-hydroxylated D3 enhance human osteoblast differentiation and mineralization. Journal of Cellular Biochemistry.

2006;**99**(3):922-935. DOI: 10.1002/jcb.20875

[28] Dauletbaev N, Herscovitch K, Das M, Chen H, Bernier J, Matouk E, et al. Down-regulation of IL-8 by high-dose vitamin D is specific to hyperinflammatory macrophages and involves mechanisms beyond up-regulation of DUSP1. British Journal of Pharmacology. 2015;**172**(19):4757-4771. DOI: 10.1111/bph.13249

[29] Dominguez LJ, Veronese N, Guerrero-Romero F, Barbagallo M. Magnesium in infectious diseases in older people. Nutrients. 2021;**13**(1):180. DOI: 10.3390/nu13010180

[30] García IM, Altamirano L, Mazzei L, Fornés M, Molina MN, Ferder L, et al. Role of mitochondria in paricalcitol-mediated cytoprotection during obstructive nephropathy. American Journal of Physiology. Renal Physiology. 2012;**302**:1595-1605. DOI: 10.1152/ajprenal.00617.2011.-Vitamin

[31] Silvagno F, de Vivo E, Attanasio A, Gallo V, Mazzucco G, Pescarmona G. Mitochondrial localization of vitamin D receptor in human platelets and differentiated megakaryocytes. PLoS ONE. 2010;**5**:1371

[32] Cucalón Arenal JM, Blay Cortés MG, Zumeta Fustero J, Blay Cortés V. Actualización en el tratamiento con colecalciferol en la hipovitaminosis D desde atención primaria. Medicina General y de Familia. 2019;**8**:68

[33] Segovia-Mendoza M, García-Quiroz J, Díaz L, García-Becerra R. Combinations of calcitriol with anticancer treatments for breast cancer: An update. International Journal of Molecular Sciences. 2021;**22**(23):12741. DOI: 10.3390/ijms222312741

[34] Castellano-Castillo D, Morcillo S, Clemente-Postigo M, Crujeiras AB, Fernandez-García JC, Torres E, et al. Adipose tissue inflammation and VDR expression and methylation in colorectal cancer. Clinical Epigenetics. 2018;**10**:60. DOI: 10.1186/s13148-018-0493-0

[35] Lei M, Liu Z, Guo J. The emerging role of Vitamin D and Vitamin D receptor in diabetic nephropathy. BioMed Research International. 2020;**2020**:4137268. DOI: 10.1155/2020/4137268

[36] Santos JM, Hussain F. VD3 mitigates breast cancer aggressiveness by targeting V-H+-ATPase. The Journal of Nutritional Biochemistry. 2019;**70**:185-193. DOI: 10.1016/j.jnutbio.2019.05.005

[37] Giulietti A, van Etten E, Overbergh L, Stoffels K, Bouillon R, Mathieu C. Monocytes from type 2 diabetic patients have a pro-inflammatory profile. 1,25-Dihydroxyvitamin D3 works as anti-inflammatory. Diabetes Research and Clinical Practice. 2007;**77**(1):47-57. DOI: 10.1016/j.diabres.2006.10.007

[38] Li J, Li B, Jiang Q, et al. Do genetic polymorphisms of the vitamin D receptor contribute to breast/ovarian cancer? A systematic review and network meta-analysis. Gene. 2018;**677**:211-227. DOI: 10.1016/j.gene.2018.07.070

[39] Caccamo D, Cannata A, Ricca S, Catalano LM, Montalto AF, Alibrandi A, et al. Role of Vitamin-D Receptor (VDR) single nucleotide polymorphisms in gestational hypertension development: A case-control study. PLoS One. 2020;**15**(11):e0239407. DOI: 10.1371/journal.pone.0239407

[40] Bronner F. Recent developments in intestinal calcium absorption. Nutrition Reviews. 2009;**67**(2):109-113. DOI: 10.1111/j.1753-4887.2008.00147.x

[41] Lucisano S, Buemi M, Passantino A, Aloisi C, Cernaro V, Santoro D. New insights on the role of vitamin D in the progression of renal damage. Kidney & Blood Pressure Research. 2013;**37**(6): 667-678. DOI: 10.1159/000355747

[42] Dominguez LJ, Farruggia M, Veronese N, Barbagallo M. Vitamin d sources, metabolism, and deficiency: Available compounds and guidelines for its treatment. Metabolites. 2021;**11**(4):3390

[43] Czernichow S, Fan T, Nocea G, Sen SS. Calcium and vitamin D intake by postmenopausal women with osteoporosis in France. Current Medical Research and Opinion. 2010;**26**(7):1667-1674. DOI: 10.1185/03007995.2010.483658

[44] Hilger J, Friedel A, Herr R, Rausch T, Roos F, Wahl DA, et al. A systematic review of vitamin D status in populations worldwide. The British Journal of Nutrition. 2014;**111**(1):23-45. DOI: 10.1017/S0007114513001840

[45] Grønli O, Kvamme JM, Jorde R, Wynn R. Vitamin D deficiency is common in psychogeriatric patients, independent of diagnosis. BMC Psychiatry. 2014;**14**:134. DOI: 10.1186/1471-244X-14-134

[46] Jones KS, Meadows SR, Schoenmakers I, Prentice A, Moore SE. Vitamin D status increases during pregnancy and in response to Vitamin D supplementation in Rural Gambian Women. The Journal of Nutrition. 2020;**150**(3):492-504. DOI: 10.1093/jn/nxz290

[47] Gröber U, Kisters K. Influence of drugs on vitamin D and calcium metabolism. Dermato-Endocrinology. 2012;**4**(2):158-166. DOI: 10.4161/derm.20731

[48] Ruiz-Ojeda FJ, Anguita-Ruiz A, Leis R, Aguilera CM. Genetic factors and molecular mechanisms of Vitamin D and obesity relationship. Annals of Nutrition & Metabolism. 2018;**73**(2):89-99. DOI: 10.1159/000490669

[49] Pereira-Santos M, Costa PR, Assis AM, Santos CA, Santos DB. Obesity and vitamin D deficiency: A systematic review and meta-analysis. Obesity Reviews. 2015;**16**(4):341-349. DOI: 10.1111/obr.12239

[50] Akkawi I, Zmerly H. Osteoporosis: Current concepts. Joints. 2018;**6**(2):122-127. DOI: 10.1055/s-0038-1660790

[51] Sozen T, Ozisik L, Calik BN. An overview and management of osteoporosis. European Journal of Rheumatology. 2017;**4**(1):46-56. DOI: 10.5152/eurjrheum.2016.048

[52] Hansen KE. High-dose vitamin D: Helpful or harmful? Current Rheumatology Reports. 2011;**13**(3): 257-264. DOI: 10.1007/s11926-011-0175-9

[53] Camacho PM, Petak SM, Binkley N, Diab DL, Eldeiry LS, Farooki A, et al. American association of clinical endocrinologists/American college of endocrinology clinical practice guidelines for the diagnosis and treatment of postmenopausal osteoporosis-2020 update. Endocrine Practice. 2020;**26**(s1):1-46. DOI: 10.4158/GL-2020-0524SUPPL

[54] Schöttker B, Herder C, Rothenbacher D, Perna L, Müller H, Brenner H. Serum 25-hydroxyvitamin D levels and incident diabetes mellitus type 2: A competing risk analysis in a large population-based cohort of older adults. European Journal of Epidemiology. 2013;**28**(3):267-275. DOI: 10.1007/s10654-013-9769-z

[55] Wolden-Kirk H, Overbergh L, Christesen HT, Brusgaard K, Mathieu C. Vitamin D and diabetes: Its importance for beta cell and immune function. Molecular and Cellular Endocrinology. 2011;**347**(1-2):106-120. DOI: 10.1016/j.mce.2011.08.016

[56] Mirhosseini N, Rainsbury J, Kimball SM. Vitamin D Supplementation, Serum 25(OH)D Concentrations and Cardiovascular Disease Risk Factors: A Systematic Review and Meta-Analysis. Frontiers in Cardiovascular Medicine. 2018;**5**. DOI: 10.3389/fcvm.2018.00087

[57] Zilberman JM. Menopausia: Hipertension arterial y enfermedad vascular. Hipertension y Riesgo Vascular. 2018;**35**(2):77-83. DOI: 10.1016/j.hipert.2017.11.001

[58] Wood AD, Secombes KR, Thies F, Aucott L, Black AJ, Mavroeidi A, et al. Vitamin D 3 supplementation has no effect on conventional cardiovascular risk factors: A parallel-group, double-blind, placebo-controlled RCT. The Journal of Clinical Endocrinology and Metabolism. 2012;**97**(10):3557-3567. DOI: 10.1210/jc.2012-2126

[59] Khera A, Kanta P, Kalra J, Dumir D, Thungapathra M. Resveratrol restores the level of key inflammatory cytokines and RANKL/OPG ratio in the femur of rat osteoporosis model. Journal of Women & Aging. 2019;**31**(6):540-552. DOI: 10.1080/08952841.2018.1522126

[60] Sato T, Watanabe K, Masuhara M, Hada N, Hakeda Y. Production of IL-7 is increased in ovariectomized mice, but not RANKL mRNA expression by osteoblasts/stromal cells in bone, and IL-7 enhances generation of osteoclast precursors in vitro. Journal of Bone and Mineral Metabolism. 2007;**25**(1):19-27. DOI: 10.1007/s00774-006-0723-y

[61] Azizieh FY et al. Circulatory pattern of cytokines, adipokines and bone markers in postmenopausal women with low BMD. Journal of Inflammation Research. 2019;**12**:99-108. DOI: 10.2147/JIR.S203590

[62] Cioffi M, Esposito K, Vietri MT, Gazzerro P, D'Auria A, Ardovino I, et al. Cytokine pattern in postmenopause. Maturitas. 2002;**41**(3):187-192. DOI: 10.1016/s0378-5122(01)00286-9

[63] Deselm CJ, Takahata Y, Warren J, Chappel JC, Khan T, Li X, et al. IL-17 mediates estrogen-deficient osteoporosis in an Act1-dependent manner. Journal of Cellular Biochemistry. 2012;**113**(9):2895-2902. DOI: 10.1002/jcb.24165

[64] Yasui T, Maegawa M, Tomita J, Miyatani Y, Yamada M, Uemura H, et al. Changes in serum cytokine concentrations during the menopausal transition. Maturitas. 2007;**56**(4):396-403. DOI: 10.1016/j.maturitas.2006.11.002

[65] Yasui T, Maegawa M, Tomita J, Miyatani Y, Yamada M, Uemura H, et al. Association of serum cytokine concentrations with psychological symptoms in midlife women. Journal of Reproductive Immunology. 2007;**75**(1):56-62. DOI: 10.1016/j.jri.2007.02.004

[66] Malutan AM, Dan M, Nicolae C, Carmen M. Proinflammatory and anti-inflammatory cytokine changes related to menopause. Przeglad Menopauzalny. 2014;**13**(3):162-168. DOI: 10.5114/pm.2014.43818

[67] Yasui T, Uemura H, Yamada M, Matsuzaki T, Tsuchiya N, Noguchi M, et al. Associations of interleukin-6 with interleukin-1β, interleukin-8 and macrophage inflammatory protein-1β in midlife women. Cytokine. 2008;**41**(3):302-306. DOI: 10.1016/j.cyto.2007.12.001

[68] Wu N, Wang QP, Li H, Wu XP, Sun ZQ, Luo XH. Relationships between serum adiponectin, leptin concentrations and bone mineral density, and bone biochemical markers in Chinese women. Clinica Chimica Acta. 2010;**411**(9-10):771-775. DOI: 10.1016/j.cca.2010.02.064

[69] Miyatani Y, Yasui T, Uemura H, Yamada M, Matsuzaki T, Kuwahara A, et al. Associations of circulating adiponectin with estradiol and monocyte chemotactic protein-1 in postmenopausal women. Menopause. 2008;**15**(3):536-541. DOI: 10.1097/gme.0b013e31815c85ed

[70] Jü J, Jü T. Plasma adiponectin concentration in healthy pre-and postmenopausal women: Relationship with body composition, bone mineral, and metabolic variables. American Journal of Physiology. Endocrinology and Metabolism. 2007;**293**:42-47. DOI: 10.1152/ajpendo.00610.2006.-The

[71] Tyagi AM, Srivastava K, MansooriMN,TrivediR,ChattopadhyayN, Singh D. Estrogen deficiency induces the differentiation of IL-17 secreting Th17 cells: A new candidate in the pathogenesis of osteoporosis. PLoS ONE. 2012;7:1371

[72] Shukla P, Mansoori MN, Singh D. Efficacy of anti-IL-23 monotherapy versus combination therapy with anti-IL-17 in estrogen deficiency induced bone loss conditions. Bone. 2018;**110**:84-95. DOI: 10.1016/j.bone.2018.01.027

[73] Tyagi AM, Mansoori MN, Srivastava K, Khan MP, Kureel J, Dixit M, et al. Enhanced immunoprotective effects by anti-il-17 antibody translates to improved skeletal parameters under estrogen deficiency compared with anti-RANKL and anti-TNF-α antibodies. Journal of Bone and Mineral Research. 2014;**29**(9):1981-1992. DOI: 10.1002/jbmr.2228

[74] Barone A, Giusti A, Pioli G, Girasole G, Razzano M, Pizzonia M, et al. Secondary hyperparathyroidism due to hypovitaminosis D affects bone mineral density response to alendronate in elderly women with osteoporosis: A randomized controlled trial. Journal of the American Geriatrics Society. 2007;**55**(5):752-757. DOI: 10.1111/j.1532-5415.2007.01161.x

[75] Neuprez A, Bruyère O, Collette J, Reginster JY. Vitamin D inadequacy in Belgian postmenopausal osteoporotic women. BMC Public Health. 2007;7:1471

[76] Holick MF, Siris ES, Binkley N, Beard MK, Khan A, Katzer JT, et al. Prevalence of vitamin D inadequacy among postmenopausal North American women receiving osteoporosis therapy. The Journal of Clinical Endocrinology and Metabolism. 2005;**90**(6):3215-3224. DOI: 10.1210/jc.2004-2364

[77] Jeffery LE, Wood AM, Qureshi OS, Hou TZ, Gardner D, Briggs Z, et al. Availability of 25-Hydroxyvitamin D 3 to APCs Controls the Balance between Regulatory and Inflammatory T Cell Responses. The Journal of Immunology. 2012;**189**(11):5155-5164. DOI: 10.4049/jimmunol.1200786

[78] Small AG, Harvey S, Kaur J, Putty T, Quach A, Munawara U, et al. Vitamin D upregulates the macrophage complement receptor immunoglobulin in innate immunity to microbial pathogens. Community Biology. 2021;**4**(1):401. DOI: 10.1038/s42003-021-01943-3

[79] Janeway CA Jr, Bottomly K. Signals and signs for lymphocyte responses. Cell. 1994;**76**(2):275-285. DOI: 10.1016/0092-8674(94)90335-2

[80] Lopez DV, Al-Jaberi FAH, Woetmann A, Ødum N, Bonefeld CM, Kongsbak-Wismann M, et al. Macrophages control the bioavailability

of Vitamin D and Vitamin D-regulated T cell responses. Frontiers in Immunology. 2021;**12**:722806. DOI: 10.3389/fimmu.2021.722806

[81] Rao Muvva J, Parasa VR, Lerm M, Svensson M, Brighenti S. Polarization of human monocyte-derived cells with Vitamin D promotes control of mycobacterium tuberculosis infection. Frontiers in Immunology. 2020;**10**:3157. DOI: 10.3389/fimmu.2019.03157

[82] Deroisy R, Collette J, Albert A, Jupsin I, Reginster JY. Administration of a supplement containing both calcium and vitamin D is more effective than calcium alone to reduce secondary hyperparathyroidism in postmenopausal women with low 25(OH)vitamin D circulating levels. Aging Clinical and Experimental Research. 2002;**14**(1):13-17. DOI: 10.1007/BF03324412

[83] McClung MR, Wagman RB, Miller PD, Wang A, Lewiecki EM. Observations following discontinuation of long-term denosumab therapy. Osteoporosis International. 2017;**28**(5):1723-1732. DOI: 10.1007/s00198-017-3919-1

[84] Tsourdi E, Langdahl B, Cohen-Solal M, Aubry-Rozier B, Eriksen EF, Guañabens N, et al. Discontinuation of Denosumab therapy for osteoporosis: A systematic review and position statement by ECTS. Bone. 2017;**105**:11-17. DOI: 10.1016/j.bone.2017.08.003

[85] Chung HY, Chin SO, Kang M. Efficacy of risedronate with cholecalciferol on 25-hydroxyvitamin D level and bone turnover in Korean patients with osteoporosis. Clinical Endocrinology. 2011;**74**(6):699-704. DOI: 10.1111/j.1365-2265.2011.04041.x

[86] Park SY, Kang M, Park HM, Rhee Y, Moon SH, Yoon HK, et al. Efficacy of risedronate with cholecalciferol on bone mineral density in Korean patients with osteoporosis. Archives of Osteoporosis. 2020;**15**(1):1165

[87] Ralston SH, Binkley N, Boonen S, Kiel DP, Reginster JY, Roux C, et al. Randomized trial of alendronate plus vitamin D3 versus standard care in osteoporotic postmenopausal women with vitamin D insufficiency. Calcified Tissue International. 2011;**88**(6):485-494. DOI: 10.1007/s00223-011-9482-4

Chapter 9

Prevention and Treatment of Diabetic Nephropathy with Vitamin D

Marijana Petrovic, Stanko Petrovic and Katarina Obrencevic

Abstract

The number of people suffering from diabetes mellitus, especially Type 2 diabetes mellitus, is increasing every year. Approximately one-third of the patients with diabetes mellitus will develop diabetic nephropathy and chronic kidney disease. Diabetic nephropathy represents the main cause of end-stage renal disease. Vitamin D deficiency is often present in patients with diabetes mellitus and could present a risk factor for a higher incidence of cardiovascular events. Vitamin D supplementation could have a renoprotective effect and the potential to delay occurrence and slow down the progression of diabetic nephropathy. The renoprotective effect is reflected in better glycoregulation, reduction of proteinuria and proinflammatory cytokines, and improved lipid regulation. New research shed the light on the important role of vitamin D in reducing renal fibrosis and stabilization of podocyte function. If we take into consideration the cost of end-stage renal disease treatment and the quality of life of patients on dialysis, any delay in end-stage renal disease is significant.

Keywords: vitamin D, diabetes mellitus, diabetes nephropathy, podocyte, urinary biomarker

1. Introduction

Diabetes mellitus (DM) is a chronic metabolic disease and one of the leading health problems in the world. The incidence is constantly increasing, not only in the developed countries of western Europe and America but also in the developing countries due to abrupt changes in lifestyle [1, 2]. According to the 2019 Global Burden of Disease project, DM is the fifth leading cause of death and the eighth leading cause of disability worldwide, with an increasing prevalence in low-middle income countries (LMICs) [1]. In addition, LMICs have a higher percentage of undiagnosed diabetes than in high-income countries (80–90% in sub-Saharan Africa versus 20–30% in western Europe and North America) [1].

According to the International Diabetes Federation (IDF) data for 2021, there are 537 million adults (aged 20–79) living with DM in the world. This number is projected to rise to 643 million by 2030 and 783 million by 2045. More than 3 out of 4 adults with DM live in LMICs. In 2021, DM is responsible for 6.7 million deaths in adults aged 20–79 years, excluding the risk of mortality associated with the COVID-19 pandemic. This corresponds to 12.2% of the global deaths of all causes in this age group. Of course, the percentage of mortality varies across different regions. Approximately one-third (32.6%) of all diabetic deaths are people of working age (under 60). This corresponds to 11.8% of the total number of global deaths in people under the age of 60. It is estimated that one person dies of DM every 5 seconds [3].

The cost of treating DM patients and its complications is very high. During 2021, at least $966 billion was spent on DM therapy, which represents an increase of 316% in the last 15 years [3].

About 541 million adults have impaired glucose tolerance (IGT) and therefore a high risk of developing Type 2 diabetes mellitus (T2DM). Life expectancy prolongation as well as the global increase in obesity strongly affects the increase in the prevalence of T2DM [3–5].

The prevalence of DM is similar between the sexes. Although according to IDF data, the estimated prevalence of DM in men, aged 20–79 years, is slightly higher than in women (10.8% vs 10.2%). Only at the age above 70 years, the percentage of female patients is higher, presumably due to the longer women's lifespan [3].

About 87–91% of DM patients have T2DM, which is also caused by the growing epidemic of obesity worldwide. The frequency depends on genetic predisposition and environmental factors. Risk factors for the development of T2DM are obesity, a positive family history, belonging to certain ethnic groups, female sex, low body weight at birth, excessive weight in the adolescent period, and gestational diabetes of the mother during pregnancy increases the risk of the newborn developing DM in the later period. An increase in the incidence of T2DM is found in an increasingly young population [1, 3, 6].

The incidence of type 1 DM (T1DM) is also on the rise, and the lower age limit is constantly lowering, most commonly occurring in Europe and North America [3]. Risk factors for T1DM are genetic predisposition, viral infections, especially enterovirus infections, immunization (only in genetically predisposed), diet, exposure to cow's milk in early life, higher socioeconomic status, obesity, vitamin D deficiency, and perinatal factors, such as maternal age, preeclampsia history, and neonatal jaundice. Contrary to T2DM, low birth weight reduces the risk of developing T1DM [3, 7].

A lot of complications develop during DM. Besides the changes in large blood vessels, small blood vessels are also affected called microvascular complications. Microvascular complications arise from changes in the arterioles, capillaries, and venules, which are responsible for the control of the permeability of blood vessels and the muscle tone that adjusts blood flow to the metabolic needs of tissues. DM causes pathognomonic changes in small blood vessels, affecting the basal membrane of arterioles in the glomerulus, retina, myocardium, skin, and muscles, leading to its thickening [8]. One of the most common microvascular complications is diabetic nephropathy (DN) or diabetic kidney disease, which develops in just over a third of DM patients. Diabetic nephropathy is the leading cause of end-stage renal disease (ESRD) in the world [9, 10].

2. Diabetic nephropathy

DN is a chronic microvascular complication of DM, manifested as a clinical syndrome characterized by persistent albuminuria (urine albumin-to-creatinine ratio (UACR) > 300 mg/g), progressive reduction in glomerular filtration rate (GFR), hypertension, and an increase in the incidence of cardiovascular events and cardiovascular fatal events [11]. In addition to the term diabetic nephropathy, Dr. Krolewski introduced the term diabetic kidney disease in 1995 to indicate clinically diagnosed kidney disease caused by DM, and the term DN should be reserved for pathohistologically confirmed kidney damage by DM; the same attitude is expressed by the Japanese Society of Pathology and the Japanese Society of Nephrology [12]. Both terms are still used in the literature. Based on previous research, we distinguish two forms of diabetic nephropathy: proteinuric diabetic nephropathy (P-DN) and non-proteinuric diabetic nephropathy (NP-DN) [13].

2.1 Pathophysiological mechanisms in diabetic nephropathy

The two main risk factors for developing DN are poor glyco-regulation and duration of DM. The mechanisms by which hyperglycaemia causes DN are complex and numerous, with renal haemodynamics disorder, impaired glucose metabolism, ischaemia, increased oxidative stress, inflammation, and increased activity of the renin-angiotensin-aldosterone system (RAAS) at the kidney level being the most important [14].

When glycoregulation is normal, 90% of the glucose filtered in the glomeruli is reabsorbed in the S1 segment of the proximal tubule via the sodium-glucose cotransporter-2 (SGLT2) receptors on the brush epithelium, and only 10% via the sodium-glucose cotransporter-1 (SGLT1) receptors located in the distal parts of the proximal tubules. In hyperglycaemia, a large amount of glucose is filtered in the glomeruli resulting in hypertrophy of the proximal tubule due to an attempt of the proximal tubule to reabsorb this glucose. The absorption of glucose is associated with sodium chloride absorption via the SGLT2 receptor. Due to the hypertrophy of the proximal tubule glucose and sodium chloride, reabsorption are increased, resulting in a smaller amount of sodium chloride arriving in the distal tubule, which is recognized by macula densa as hypovolemia and consequently reduces tubuloglomerular feedback (TGF). TGF is the mechanism by which the macula densa controls the vasoconstriction of the afferent arteriole by releasing adenosine, and other signalling molecules. By blocking TGF, vasodilatation of the afferent arteriole occurs via vasoactive mediators, such as insulin-like growth factor-1 (IGF-1), nitric oxide (NO), prostaglandins, and glucagon. At the same time, due to the increased local effect of angiotensin II in hyperglycaemia, increased vasoconstriction of the efferent arteriole and the consequent disruption of autoregulation occur, resulting in the development of intraglomerular hypertension. Thus, the "sodium paradox" is expressed in DN, the higher the sodium chloride restriction in the diet, the smaller the amount of sodium chloride that arrives in the distal tubule and the GFR increases. GFR may be increased up to 40%. In normoglycemia, changes in systemic perfusion pressure are not transferred to intraglomerular pressure due to autoregulation, but since autoregulation is compromised in DM, systemic perfusion pressure is freely transferred to the capillary glomerular network [14, 15].

Functional changes in blood flow and glomerular pressure provoke structural changes. Increased intraglomerular pressure damages the endothelial surface of glomerular capillaries, disrupts the normal structure of the glomerular barrier, and induces the expression of cytokines and growth factors that cause the increased synthesis of collagen, fibronectin, and laminin. Hyperfiltration is associated with glomerulomegaly and renal enlargement. Glomerular enlargement is due to an increase in the number of capillary loops with a consequent increase in the filtration area and enlargement of the mesangial matrix [16].

Changes also affect tubular cells, besides glomeruli, such as proliferation and increase in cellular size. Mechanisms by which increased blood glucose levels induce hypertrophy to include stimulation and increased expression of multiple renal growth factors, including IGF-1, epidermal growth factor, platelet-derived growth factor (PDGF), vascular endothelial growth factor (VEGF), and transforming growth factor-Beta 1 (TGF-β1) [16–18].

The following section will briefly describe the main pathogenetic mechanisms in DN.

2.1.1 The role of protein kinase C

One of the main side effects of hyperglycaemia is the increased activation of protein kinase C (PKC). PKC is an enzyme, a member of the serine-threonine-protein kinase family that regulates various vascular functions, including contractility, blood flow, cellular proliferation, and vascular permeability. PKC activity in DM patients is particularly increased on the surface of the retina, aorta, heart, and glomerular cells including podocytes, presumably due to increased de novo synthesis of diacylglycerol, being the main endogenous PKC activator [19, 20].

Several isoforms of PKC exist, however, PKCα is considered to be the most important for DN development. The main role of PKCα is in increased endocytosis of nephrin, which is one of the main components of the slit diaphragm (SD). Increased expression of PKCα on podocytes in DM patients increases nephrin endocytosis leading to SD instability [20, 21]. The PKCβ isoform is responsible for hyperglycaemia-induced renal fibrosis via the upstream regulation of TGF-β1, which contributes to the development of glomerular fibrosis. The PKC is form activated by advanced glycation end products (AGEs) leads to increased expression of VEGF164 and VEGF165 isoforms at the podocytes, which leads to structural and functional renal changes seen in animal models of DN, including proteinuria, glomerular hypertrophy, thickening of the glomerular basal membrane, mesangium proliferation, and loss of SD and foot extensions of podocytes [18–20].

2.1.2 Effect of advanced glycation end products

Chronic hyperglycaemia leads to non-enzymatic glycosylation of amino acids and proteins known as AGEs. AGEs mediate various cellular activities, including molecule expression and adhesion, cellular hypertrophy, extracellular matrix synthesis, epithelial-mesenchymal transformation of tubular cells, and inhibition of nitrogen monoxide synthesis. They can bind to a variety of cell types, including macrophages and mesangial cells. Binding to macrophages leads to increased production and release of proinflammatory mediators: tumor necrosis factor alpha (TNF-α), interleukins, PDGF, and IGF-1. Released mediators stimulate mesangial cells to increase mesangial matrix synthesis, which contributes to glomerulosclerosis. AGEs stimulate proximal tubule epithelial cells to increase TGF-β1 synthesis and its release, leading to tubulointerstitial fibrosis [18, 20, 22].

2.1.3 Effect of sorbitol

Hyperglycaemia stimulates the polyol pathway of glucose metabolism and increased sorbitol production. Chronic hyperglycaemia leads to increased glucose uptake in tissues that do not require insulin for glucose uptake, such as retina, eye lens and kidney. Increased glucose concentration in the tissue leads to sorbitol accumulation. The deposition of sorbitol in the epithelial cells of the proximal tubules leads to decreased free myo-inositol and decreased activity of sodium potassium adenosine triphosphatase (Na+/K+-ATP-ase), an increase in the consumption of the nicotinamide dinucleotide phosphate (NADPH) and nicotinamide dinucleotide (NAD+), which results in a change in cell oxidative potential and disruption of cell function [18, 20, 22].

2.1.4 Renin-angiotensin-aldosterone system in diabetic nephropathy

Local RAAS activity is increased in DM patients, that is, activity is increased in the glomerulus and renal vessels. Prorenin production is increased and prorenin binds to specific prorenin-renin receptors, followed by non-enzymatic activation of prorenin into renin, then renin induces local production of angiotensin II. The expression of angiotensin II type 1 receptor (AT1R) is increased, through which angiotensin II, as a potent vasoconstrictor, induces vasoconstriction of the efferent arteriole, which, in addition to dilatation of the afferent arteriole, leads to increased hydrostatic pressure in the glomerular capillaries and increased glomerular filtration. Increased activity of angiotensin II also leads to an increase in systemic blood pressure. Angiotensin II also directly stimulates renal production of TGF-β1 in the mesangium and epithelial tubular cells and stimulates the production of other cytokines and growth factors in renal cells, such as endothelin-1, monocyte chemoattractant protein-1 (MCP-1), interleukin-6 (IL-6), regulated upon activation, normal T cell expressed and presumably secreted (RANTES), and osteopontin. In addition to lowering intraglomerular pressure, angiotensin-converting enzyme inhibitors (ACEi) reduce renal expression of TGF-β1. Angiotensin II increases the release of VEGF from podocytes, which contributes to the disruption of the filtration barrier and the progression of proteinuria. All released mediators contribute to the development of glomerulosclerosis and tubulointerstitium fibrosis. The aldosterone release is increased, which also accelerates the progression of renal damage independent of angiotensin II. Several clinical studies have shown that blocking the mineral corticosteroid receptor lowers proteinuria [14, 18, 22].

2.1.5 Inflammation and diabetic nephropathy

DN was previously defined as a non-inflammatory disease. However, the results of numerous clinical studies show that activation of the immune system and chronic inflammatory processes play a very important role in the development of DN [20, 23].

The most significant inflammatory cells involved in the development of DN are macrophages (M1 and M2 macrophages) and T lymphocytes. M1 macrophages activated by Th1 cells increase the inflammatory response by increased expression of cytokines (interleukins, TNF-α, and interferon γ), unlike M2 macrophages (activated by Th2 cells), which lead to increased expression of anti-inflammatory cytokines. M1 macrophages increase the inflammatory response also by increasing the production of free oxygen radicals [14, 18, 23].

Numerous chemokines are involved in the inflammatory response in DN, the most significant and first described is MCP-1, which also plays an important role in the early development of atherosclerosis. MCP-1 induces the transformation of monocytes into macrophages, which then increasingly produce IL-6 and TNF-α, leading to initial atherosclerotic changes in the blood vessel wall, resulting in disease progression [14, 18, 20, 23].

In DM patients, the secretion of proinflammatory cytokines, interleukin-1 (IL-1), interleukin-6 (IL-6), and interleukin-18 (IL-18), is increased. IL-1 induces increased expression of adhesion molecules in glomerular endothelial cells and other renal structures, leading to changes in glomerular haemodynamics, vascular permeability, and increased chemokine expression, resulting in proliferation and synthesis of extracellular matrix in the mesangium. IL-6 in DN has pleiotropic effects, the proliferation of the extracellular matrix, thickening of the glomerular membrane, and is considered important for the progression of DN. Local IL-18 in the kidney DM patients is secreted by T lymphocytes, macrophages, monocytes, and proximal tubule cells. It increases the expression of intracellular adhesion molecule 1 (ICAM-1), as well as the production of other inflammatory cytokines in mesangial cells, and is responsible for endothelial apoptosis. A direct correlation between IL-18 and increased urinary excretion of albumin has been confirmed, some authors even consider it as an early indicator of DN [14, 18, 20, 23].

TNF-α has multiple roles in the inflammatory response, and it is synthesized by monocytes, macrophages, and T lymphocytes, as well as by some renal cells. It leads to the activation of secondary messengers, transcriptional factors, growth factors, adhesion molecules, and the expression or synthesis of other cytokines. TNF-α as well as the before mentioned interleukins in renal structures changes the haemodynamics within glomeruli, increases vascular permeability, increases infiltration by inflammatory cells, and increases extracellular matrix production and production of reactive oxygen species (ROS) [14, 18, 20, 23].

Adhesion molecules ICAM-1 and vascular cell adhesion molecule-1 (VCAM-1) allow the leukocytes to bind to the vascular wall and their infiltration of tunica intima, where they secrete proteolytic enzymes responsible for the tissue damage and the development of atherosclerosis [18].

2.1.6 Oxidative stress in diabetic nephropathy

Oxidative stress plays a significant role in the development of microvascular complications of DM. Metabolic abnormalities in DM, and above all hyperglycaemia leads to increased production of ROS in mitochondria. Increased production of ROS at the mitochondrial level causes activation of the five main pathways involved in the development of microvascular complications of DM: polyol pathway, increased AGEs production, increased AGEs receptor expression and ligand activation, PKC isoform activation, and increased hexosamine pathway activity. Through these pathways, the increased concentration of intracellular ROS causes a change in angiogenesis in response to ischemia, activates numerous proinflammatory pathways, and causes long-acting epigenetic changes. These epigenetic changes lead to persistent expression of proinflammatory genes also after normalization of blood glucose (so-called hyperglycaemic memory) [14, 18, 20].

Persistent expression of proinflammatory genes is achieved through multiple signalling pathways; however, the importance of the JAK/STAT pathway is particularly emphasized [20, 24].

2.2 Diagnosis and clinical manifestation of diabetic nephropathy

2.2.1 Diagnosis and clinical manifestation of proteinuric diabetic nephropathy

Diagnostic criteria for proteinuric diabetic nephropathy (P-DN) include UACR of 30 mg/g or greater [11]. According to the recommendations of the American Diabetes Association (ADA) and the National Kidney Foundation Kidney Disease Outcomes Quality Initiative (NKF KDOQI), a positive finding in at least two out of three urine samples, 3–6 months apart, is required for the diagnosis of persistent microalbuminuria [25]. Albuminuria can be determined in a spot sample of morning urine from which the UACR is calculated. UACR for normal albuminuria is < 30 mg/g, for microalbuminuria UACR is between 30 and 300 mg/g, for macroalbuminuria UACR is > 300 mg/g. Albuminuria can be calculated also from a 24-hour urine sample, where albuminuria up to 150 mg/24h is considered normoalbuminuria, microalbuminuria from 150 to 300 mg/24h, and macroalbuminuria over 300 mg/24h [11].

DN develops through five different phases: 1st phase glomerular hyperfiltration, 2nd phase normoalbuminuria, 3rd phase microalbuminuria, 4th phase macroalbuminuria, and 5th phase renal insufficiency (GFR less than 60 mL/minute/1.73 m^2). These five phases are clearly separated in T1DM, while in T2DM when diagnosed, microalbuminuria may be present, due to the long silent lasting of hyperglycaemia [11]. Changes in the first three phases are reversible, especially in patients with T2DM. Japanese authors reported regression of microalbuminuria in almost 50% of the patients followed [26]. Japanese and other authors published that regression was positively correlated with glycated haemoglobin A1c (HbA1c) values, systolic blood pressure, and duration of microalbuminuria [26, 27].

The occurrence of microalbuminuria and macroalbuminuria in addition to kidney damage is an independent risk factor for cardiovascular diseases. P-DN is developed in patients with T1DM and T2DM. In patients with T1DM, DN occurs after a longer disease evolution, over 10 years, and is usually associated with diabetic retinopathy (DR). In T2DM, DN can be diagnosed immediately after DM diagnosis, even without DR [11, 13].

2.2.2 Diagnosis and clinical manifestation of non-proteinuric diabetic nephropathy

Some DM patients have a reduction in GFR of less than 60 mL/minute/1.73 m^2 without microalbuminuria. This type of DN is called non-proteinuric diabetic nephropathy (NP-DN). NP-DN is more commonly associated with female sex, hypertension, poor glyco-regulation, smoking, and use of RAAS blockers. It occurs more often in patients with T2DM. NP-DN shows a weak correlation with retinopathy but shows a linear correlation with the percentage of cardiovascular complications, that is, with macroangiopathy [12, 13, 28].

According to the results of the DEMAND study in the population of patients with T2DM and creatinine clearance lower than 60 ml/min/1.73m2, about 40% of patients had normal albumin secretion. These individuals were shown to have a higher degree of insulin resistance, a higher level of serum triglycerides and total cholesterol, and low-density lipoprotein (LDL), that is, a more pronounced metabolic syndrome compared to those with preserved GFR [29]. In his work, Krolewski proposes the theory that NP-DN occurs primarily as part of inflammation and tubular lesions, while

Reutens believes that repeated and untreated episodes of acute renal impairment play a key role in NP-DN [30, 31]. Authors published that endothelial dysfunction and increased vascular resistance in interlobar renal arteries are important for the pathogenesis of NP-DN, which can be monitored by measuring the renal resistance index if elevated in patients with DM a more intensive treatment of DM should be started even before decreasing GFR [20, 32, 33].

Based on several studies, it is concluded that patients with NP-DN compared to P-DN still have a slower GFR decrease, better blood pressure control, and a lower risk of developing ESRD [13, 27, 28].

2.3 Pathohistological changes in diabetic nephropathy

Pathohistological changes of DN are thickening of the glomerular basal membrane, mesangium proliferation, glomerulomegaly, nodular glomerular sclerosis (Kimmelstiel-Wilson nodule), hyalinosis lesions characterized by exudative/insudative lesion and fibrin cap, arteriolar hyalinosis, and reduction in podocyte count with loss of foot extensions. Nodular glomerular sclerosis is characteristic of PN-DN. Tubulointerstitial basal membrane thickening, tubular atrophy, interstitial peritubular fibrosis, and atherosclerosis occur at the tubulointerstitium level.

Tervaret et al. proposed a classification of morphological changes in DN and divided into IV classes:

Class I: Thickening of the glomerular basal membrane—changes in GBM are present with the absence of mesangium expansion, as well as nodular changes in the mesangial matrix.

Class II: Mesangium proliferation—defined as the increase of the mesangium extracellular matrix so that the width of the interspace reaches two nuclei of mesangium cells in at least two glomeruli. Mesangium proliferation may be mild (IIa) and severe (IIb).

Class III: Nodular sclerosis (Kimmelstiel-Wilson nodes)—focal, oval, or round mesangium lesions composed of an acellular hyalinized nucleus that is surrounded by rare crescent mesangial nuclei. It is sufficient to see one clear Kimmelstiel-Wilson change in the preparation and if there is no more than 50% global glomerulosclerosis, the changes are classified as Type III.

Class IV: Global (diffuse) diabetic glomerulosclerosis—the final stage characterized by the accumulation of extracellular matrix proteins in the mesangium space with the formation of numerous Kimmelstiel-Wilson changes [34].

These changes can be seen in PN-DN and NP-DN. However, in NP-DN, changes in glomerulus are less pronounced, smaller number of patients have changes characteristic of class III and class IV, while changes in tubules have similar prevalence in both forms of DN [13, 28].

3. Vitamin D and diabetic nephropathy

Considering the burden of constantly increasing DM, with more than a third of the patients developing DN, and the fact that DN is the leading cause of ESRD, it is important to find other possible mechanisms and risk factors for the development of DN. Since the costs of ESRD treatment by haemodialysis or kidney transplantation are extremely high, constant research is being conducted to prevent the development of DN or slow its progression, and to find biomarkers for early diagnosis of DN.

Vitamin D deficiency is becoming pandemic and is found in different age populations. Vitamin D deficiency is not recognized and not adequately treated in many countries. A vitamin D deficiency is present in more than one billion people worldwide. The prevalence of vitamin D deficiency is high in countries around the world. In Italy, the prevalence in the general population is 17%, in Spain 33.9%, in Germany 50%, while in the UK, it is 87.1%. Vitamin D deficiency is associated with an increase in morbidity for cardiovascular, carcinogenic, metabolic—DM, immunological, psychological, and other chronic diseases [35].

For the normal function of vitamin D, the kidney plays an important role. Vitamin D is transported bound to vitamin D binding protein (VDBP). After the first hydroxylation is carried out in the liver, the resulting 25-hydroxyvitamin D3 (25(OH)D3) is transported to the kidney bound to VDBP, where it is first filtered in the glomerulus and then reabsorbed through the receptors on the brush border of the proximal tubule. In the kidney, in the presence of 1-alpha-hydroxylase (CIP27B1), the second hydroxylation is carried out to form a more polar compound 1,25-hydroxyvitamin D3 (1,25(OH)2D3), that is, calcitriol, being a physiologically active form of vitamin D, and based on its physiological effects, vitamin D is classified as a vitamin and as a hormone [36, 37].

Multiple studies have shown that the prevalence of 25(OH)D3 deficiency is significantly higher in patients with T1DM and T2DM compared to the healthy population. Complementary, a significant negative correlation has been found between the concentration of 25(OH)D3 and HBA1c ($r = -0.277$, $p < 0.0001$) [34]. Also, several studies have shown a significantly higher risk of developing T2DM in individuals with a concentration of 25(OH)D3 of $\leq$ 20 ng/mL ($\leq$50 nmol/L). Studies also showed a significantly strong positive correlation between 25(OH)D3 concentration and GFR. Patients with DM and concentration of 25(OH)D3<15ng/mL (<37,5 nmol/L) have a faster GFR reduction compared to patients with a concentration of 25(OH)D3>15 ng/mL(>37,5 nmol/L) [38, 39].

Patients with microalbuminuria and macroalbuminuria have significantly lower serum concentrations of 25(OH)D3 than patients with normoalbuminuria [40].

In lean DM patients, the risk of vitamin D deficiency is 25% higher than in the general population, while in obese DM patients, the risk is 35% higher. Females with DM more often have vitamin D deficiency, which is explained by the higher fat content in women with consequent formation of vitamin D depot in the body [35, 38].

DM patients have multiple risk factors for vitamin D deficiency. In DN, vitamin D metabolism may be disrupted due to major loss of VDBP as part of proteinuria, which may also be of nephrotic rank. Also, 25(OH)D3 filtered in glomerulus bound to VDBP can be reabsorbed to a significantly lesser extent at the level of proximal tubules due to their damage, which contributes to vitamin D deficiency [38].

Previous studies showed that vitamin D may have an anti-inflammatory, immunomodulatory, and hypoglycaemic effect and may affect later occurrence and slower progression of DN. These are renoprotective effects of vitamin D that are assumed that can delay the occurrence and slow down the development of DN

3.1 Effect of vitamin D on glycoregulation

The vitamin D prevents the occurrence of DM by increasing insulin sensitivity by stimulating and increasing the expression of insulin receptors in the skeletal muscles with the activation of the peroxisome proliferator activator receptor δ (PPAR δ), and by inhibiting RAAS (inhibitor of the insulin effect on peripheral tissues) [36]. Also, vitamin D affects increased insulin release by increasing intracellular calcium in pancreatic beta cells. Vitamin D can also have a direct effect on beta cell function, which is done by binding its active form in the blood to the vitamin D receptor (VDR), which is present in the beta cells of the pancreas. Mice without functional VDR have been shown to have impaired insulin secretion after glucose load [38, 41, 42]. Vitamin D also increases the level of osteocalcin synthesized in the osteoblasts, which in turn increases insulin synthesis in the pancreas and reduces the inflammation responsible for insulin resistance [38, 42]. It is considered that an optimal blood level between 80 and 119 nmol/L is required for vitamin D to achieve its effect on lowering insulin resistance [35, 43].

The use of vitamin D can reduce the risk of diabetes to a certain extent [39, 42, 44]. Thus, the observational study in Denmark "Copenhagen City Heart Study", conducted on 9841 non-DM patients followed for 29 years, showed that patients who had an average value of 25(OH)D <12.5 nmol/L (< 5 ng/ml) had a 22% higher risk of developing T2DM compared to patients with serum concentration of 25(OH)D ≥50 nmol/L (≥20 ng/ml), patients who had 25(OH)D ≥75 nmol/L (≥30ng/ml) had an 8% lower risk of morbidity compared to patients whose 25(OH)D level was between 50 and 75 nmol/L (20–30 ng/ml) [45].

A double-blind, randomized, placebo-controlled study of 118 patients with T2DM and vitamin D deficiency showed that 8-week combined administration of vitamin D at a dose of 50000 IU/week and 1000 mg/day of calcium was significantly more effective in reducing the levels of morning glycaemia and HbA1c compared to the effect of administration of individual components or placebo [46].

Petrovic et al. conducted a prospective cohort study on 90 patients with T2DM and vitamin D deficiency in whom vitamin D supplementation was conducted for six months (at the beginning, cholecalciferol drops supplementation was started at the dose of 20,000 IU twice a week). After two months in patients who normalized vitamin D levels, the supplementation was continued with 5000 IU of cholecalciferol twice a week for the next four months, and in patients whose vitamin D levels remained reduced, the substitution was continued with 20,000 IU twice a week. The study showed that vitamin D supplementation for six months led to a statistically significant reduction of HbA1c in all patients regardless of the degree of albuminuria [40].

3.2 Anti-inflammatory effect of vitamin D

Chronic inflammation is considered to play a central role in the development of DM and DN, and vitamin D can directly or indirectly reduce its effect [23]. The anti-inflammatory effect is reflected in the reduced release of numerous proinflammatory cytokines, such as TNFα, IL-6, IL-12, IL-18, IL-1β, interferon-gamma (INF-γ), dendritic cell differentiation blockade, inhibition of lymphocytic proliferation, inhibition of foam cell formation, reduced cholesterol uptake into macrophages, and improved regulatory development of T lymphocytes (Th1 inhibition and Th2 lymphocyte activation) or in increased release of anti-inflammatory cytokines, such as IL-10. Vitamin D also induces CD4+ and CD25+ regulatory lymphocytes that inhibit inflammation and inhibit the effects of TNF-α, ICAM, and VCAM-1 [20, 38, 47].

All of the above mechanisms affect the slower development of atherosclerotic changes and the lower incidence of adverse cardiovascular events.

3.3 Effect of vitamin D on proteinuria

Several authors reported the anti-proteinuric effect of correcting vitamin D deficiency in DM [40, 48–50]. Since proteinuria is the main treatment target in terms of slowing DN progression, this effect should be highlighted. The anti-proteinuric effect of vitamin D is done by the inhibition of RAAS in the juxtaglomerular kidney apparatus by downstream regulation of gene transcription, due to binding to the transcription factor cyclic adenosine monophosphate-response element (CRE)-binding protein and disabling renal transcription. In addition to the inhibition of renal RAAS, vitamin D lowers the expression of renin at the heart level, thus decreasing blood pressure. De Zeeuw confirmed the anti-proteinuric effect of vitamin D in the VITAL study, similar results were published by other authors [40, 49, 50]. In addition to the effect on renin, the hypotensive effect is also due to the direct effect on vascular cells and calcium metabolism. With these mechanisms, vitamin D leads to better control of blood pressure and intraglomerular pressure [38]. According to the literature, a negative correlation was found between serum vitamin D concentration and blood pressure level [51]. The antiproteinuric effect is also achieved by improving glyco-regulation, anti-fibrotic effect—lowering the activity of TGF-β1 in the urine, increasing the expression of nephrin at the kidney level, reducing levels and other growth factors that interfere with SD and podocyte function, such as VEGF-A and MCP-1 [38]. In addition, slowing fibrosis prevents and slows the progression of left ventricular hypertrophy and the development of cardiac weakness and reduces the expression of genes involved in atherosclerosis as well as vascular growth factors [38, 52].

3.4 Effect of vitamin D on lipid status

Correction of vitamin D deficiencies in patients with T2DM improves lipid status, and thus, can prevent cardiovascular complications [40, 53]. Xiao Fei Qin et al. have demonstrated that the administration of vitamin D in patients with reduced vitamin D levels and hyperlipidaemia, after use of statins and vitamin D (2000IU/day) together for six months, leads to a significant reduction in both triglycerides and cholesterol levels compared to patients who used placebo with statins [54]. PalomaMuñoz-Aguirre also achieved a significant reduction in triglyceride levels after administration of vitamin D at a dose of 4000 IU/day in postmenopausal women with T2DM [55]. Other authors have also confirmed a significant decrease in cholesterol and triglyceride levels after vitamin D supplementation in patients with T2DM and vitamin D deficiency [40]. Some authors have not confirmed any positive effect of vitamin D on lipid status [56].

Based on the above data, we conclude that correction of vitamin D deficiency in DM patients with DN has a significant renoprotective effect, which is reflected in the improvement of glyco-regulation, lowering of proteinuria, improvement of lipid status, anti-inflammatory effect, anti-fibrotic effect, inhibition of RAAS.

Since albuminuria is not always an indicator of initial changes within DN, identification of new renal biomarkers is crucial to establish early diagnosis and start early treatment, aiming to slow the progression of DN. Several urinary biomarkers have been distinguished from previous studies. Biomarkers of glomerular injury are nephrin, VEGF-A, collagen type IV, transferrin, podocalyxin, immunoglobulins, ceruloplasmin, laminin, fibronectin, and glycosaminoglycans. Biomarkers of the tubular injury are neutrophil gelatinase-associated lipocalin (NGAL), nephrin, VEGF-A, alpha 1-microglobulin, kidney injury molecule-1 (KIM-1), MCP-1, N-acetyl-β-D glucosaminidase, cystatin C, liver-type fatty acid binding protein (L-FABP), and others. Data in the literature related to these biomarkers still refer to studies conducted on a small number of patients, so it is necessary to plan studies on a larger number of patients. The following text will describe the additional renoprotective effect of vitamin D based on the effect on some of these biomarkers.

3.5 Effect of vitamin D on renal fibrosis in diabetic nephropathy

The renoprotective effect of vitamin D is also reflected in the anti-fibrotic effect because it has been proven that in DM patients, the use of vitamin D leads to a lowering of the level of TGF-β1 in the urine, and TGF-β1 is considered the main culprit for the development of renal fibrosis. Also, the administration of vitamin D lowers the increased synthesis of type 1 collagen and fibronectin, which are components of the extracellular matrix [38, 52].

TGF-β1 is a multifunctional cytokine considered to play one of the key roles in DN pathophysiology. In conditions of hyperglycaemia, TGF-β1 formation is increased in almost all renal cells, its expression is increased in both glomerulus and tubulointerstitium. Increased mRNA expression for TGF-β1 has been demonstrated in all renal tissue structures. In DM patients, in addition to hyperglycaemia, increased TGF-β1 synthesis is influenced by: AGEs, oxidative stress, intraglomerular hypertension, PKC activation, de novo diacylglycerol synthesis, and elevated levels of vasoactive substances, such as angiotensin II and endothelin. TGF-β1 stimulates the synthesis of the extracellular matrix. Receptors for TGF-β1 can be observed on all glomerular structures. In response to the action of TGF-β1, mesangium and epithelial cells increase the synthesis of proteoglycans, fibronectin, collagen, and laminin. TGF-β1, on the other hand, inhibits the synthesis of collagenases and stimulates the tissue production of metalloproteinase inhibitors, thereby reducing the activity of matrix metalloproteinases that are responsible for the degradation of the extracellular matrix. Matrix metalloproteinases consist of interstitial collagenase, stromelysin, and type IV collagenase, the most important being matrix metalloproteinase-2 (MMP-2) – produced by mesangium cells. In human renal bioptic material in patients with DN, downstream gene regulation for MMP-2 has been demonstrated. It has also been confirmed that the glycosylated matrix components are resistant to degradation. Angiotensin II directly stimulates renal production of TGF-β1 in the mesangium and epithelial tubular cells, also leads to the upstream regulation of TGF-β1 receptor expression, and increases their production and sensitivity of mesangium cells to this growth factor [38, 52, 57].

In addition to the increased synthesis of the extracellular matrix and its reduced degradation, TGF-β1 induces the transformation of tubular epithelial cells into fibroblasts, and this process is responsible for renal fibrosis as a result of persistent inflammation. Inhibition of TGF-β1 at the level of the proximal tubules is considered to have a greater effect on the reduction of albuminuria and renal fibrosis, while inhibition at the level of the glomeruli has a greater effect on the preservation of GFR [57, 58].

Inflammation-related effects of TGF-β1 are done mainly through SMADs. SMADs are a family of structurally similar intracellular proteins, some of which are signalling molecules, and some transcriptional factors that carry the extracellular signal after binding TGF-β1 to its receptor in the nucleus, and subsequently inhibit or stimulate the expression of certain genes [38].

Based on the above, we can conclude that urinary TGF-β1 can be considered as a biomarker of renal impairment in DN. The level of TGF-β1 according to some authors correlates with albuminuria and GFR decrease. Min Joeng Kim et al. and Yanian Tian et al. have demonstrated that the administration of vitamin D preparations can reduce the urine level of TGF-β1 in DM type 2 patients with consequent albuminuria reduction [59, 60]. The use of 1,25(OH)2D3 leads to downstream regulation of the TGF-β1 signalling pathway, lowering the expression of SMAD3 in the renal tissue of DM patients [41]. Also, the use of vitamin D in DM patients lowers hyperglycaemia-induced increased synthesis of fibronectin, type 1 collagen, the main components of the proliferated extracellular matrix. It is assumed that only TGF-β1 suppression at the renal level is significant for slowing the progression of DN. TGF-β1 suppression at the systemic level affects the metabolism of water and sodium leading to primary hyperaldosteronism [57].

3.6 Vitamin D and podocyte

Podocyte dysfunction is known to play one of the key roles in DN pathophysiology. It is assumed that in the future, the stabilization of podocyte function will play a key role in slowing the progression of DN.

Nephrin, podocin, and podocalyxin are structural proteins that determine the structure of SD and are responsible for its selective permeability. They can be considered as biomarkers of podocyte damage, nephrin also being a marker of proximal tubule damage. Within DN, the expression of nephrin, podocin, and podocalyxin at the podocyte level is reduced, but their urinary secretion is increased and is considered to correlate with proteinuria [38, 61].

Nephrin is a structural and signalling protein, member of the immunoglobulin family, responsible for the control of cytoskeletal architecture, the shape, and viability of podocytes. For adequate glomerular function, a certain level of nephrin expression is necessary. In patients with DM, mutations occur in the nephrin gene NPHS1 and the podocin gene NPHS2 leading to their reduced expression at the podocyte level. As a consequence, permeability dysfunction of the glomerular membrane and albuminuria occur. One of the main mechanisms in the development of proteinuria in diabetic rats is increased nephrin endocytosis, mediated by a complex consisting of PKCα, protein interacting with C kinase 1 (PICK1), and beta-arrestin 2 [21, 38, 62].

Nephrin is also a signal protein located on the surface of β cells of the pancreas. Nephrin phosphorylation mediates the release of insulin from pancreatic β cells. At the experimental level, Uchida K. et al. demonstrated that nephrin dephosphorylation occurring in the kidney in DM plays a significant role in the development of albuminuria [63].

Zhang et al. argue that the use of vitamin D preparations in DM and DN patients increases nephrin expression in podocyte culture. It has been confirmed that the administration of vitamin D also enhances podocin and podocalyxin expression on podocytes, thus stabilizing their function [38, 41].

VEGF-A belongs to the group of pro-angiogenic glycoproteins. It is essential for the survival, proliferation and differentiation of endothelial cells, podocytes, and mesangial cells, and is also considered responsible for normal glomerulogenesis. It is produced at the renal level by podocytes and epithelial cells of the proximal tubule. VEGF-A is a part of the SD and controls its permeability. In hyperglycaemia, its synthesis and release are increased. The histopathological finding of early DN stage renal biopsy shows increased expression of mRNA VEGF-A. Increased VEGF synthesis lowers the level of nephrin expression in renal tissue via the VEGFR2 receptor, inhibiting phosphorylation of nephrin and accelerating its endocytosis, causing SD permeability dysfunction and albuminuria. VEGF-A stimulates TGF-β1 activity under hyperglycaemic conditions and contributes to mesangial proliferation, fibrosis, and later glomerulosclerosis. VEGF-A also lowers nitrogen monoxide levels with consequent endothelial dysfunction. High VEGF concentrations are responsible for pathological angiogenesis in both the kidney and other tissues [38, 52, 62, 64].

The isoform of PKCs activated by AGEs in hyperglycaemia leads to increased expression in the podocytes of VEGF164 and VEGF165 isoforms that in animal models lead to structural and functional renal changes seen in DN, including proteinuria, glomerular hypertrophy, thickening of the glomerular basement membrane, mesangium proliferation, and loss of SD function and foot extensions [65].

In the period of initial and moderate renal tissue damage, the level of VEGF-A is increased, however, over time, as podocytes and proximal tubular cells damage progresses, as fibrosis is more pronounced, the level of VEGF-A decreases because the cells that secrete it are definitely damaged. It is assumed that the stabilization of podocyte function and control of the effect of VEGF-A will play a key role in slowing the progression of DN. Several authors have already presented the positive effect of vitamin D on VEGF-A in patients with DN, in terms of lowering its value after the administration of vitamin D [38, 64].

Some authors have not confirmed the positive effect of vitamin D in DM. A meta-analysis, including 35 studies, showed no positive effect of vitamin D on glucose homeostasis and DM prevention [66]. Different results on the effect of vitamin D in DM type 2 patients are assumed to be due to incoherent data, a small number of patients and a short follow-up period.

4. Administration of vitamin D in patients with diabetes mellitus and diabetic nephropathy

Based on the results of numerous researches and multicentric studies, several scientific associations such as the European Society for Clinical and Economic Aspects of Osteoporosis and Osteoarthritis, European Menopause and Andropause Society, Kidney Disease: Improving Global Outcomes Clinical Practice, or American Geriatrics Society, and American Academy of Developmental Medicine and Dentistry published a recommendation for the use of vitamin D. The individual approach to the patient was considered, that is, life in different climatic conditions in terms of exposure to the sun, followed by diet, physical activity, age, obesity and associated diseases [35, 67].

Serum concentration of calcidiol—25(OH)D is the most reliable indicator of vitamin D status in the humans. The recommended optimal blood level of 25(OH)D should be between 30–50 ng/mL or 75–125 nmol/L or even 40–60 ng/mL (100–150 nmol/L). It is necessary for vitamin D to reach a serum level of at least 32 ng/mL (80 nmol/L) in DM patients to express a positive effect [35, 42, 67–69].

Screening of 25(OH)D level should be performed in all DM patients, especially in patients who use chronic therapy that affects vitamin D metabolism and with associated diseases, such as osteoporosis, osteomalacia, hyperparathyroidism, cardiovascular diseases, malabsorption syndrome, hepatic insufficiency, recurrent respiratory tract infections, cancer, neurological diseases, autoimmune diseases, over 65 years of age, obese (Body mass index -(BMI)>30 kg/m2), people with non-traumatic fractures, musculoskeletal pain, and persons with black skin colour [35, 67–69].

In order to maintain optimal vitamin D level, a healthy lifestyle is recommended for all persons as a diet with sufficient vitamin D and calcium intake, safe controlled sun exposure, physical activity, weight control, and smoking cessation. When using vitamin D preparations, adequate calcium intake through diet is also necessary. If the diet cannot provide adequate calcium intake or there is a malabsorption problem, calcium supplements (1000 mg calcium /day) should be used [35, 67, 69].

A detailed overview of vitamin D administration is given in **Tables 1** and **2** [35, 67, 69].

For the treatment of vitamin D deficiency in adults, oral cholecalciferol (vitamin D3) is recommended. Calcifediol can be used instead of vitamin D in certain conditions, such as obesity, malabsorption, liver disease, chronic kidney disease (stage 3 or 4), and in all conditions where rapid correction of vitamin D deficiency is required. The use of calcifediol may also be helpful in patients taking drugs that interfere with the hepatic cytochrome P-450 enzyme system, including those taking glucocorticoids, anticonvulsants, anticancer drugs, or antiretroviral drugs. In certain risk groups (e.g., patients with severe malabsorption), vitamin D may also be administered parenterally [67, 70, 71].

Calcitriol (1,25(OH)2D) and its analogues are used at much lower doses compared to vitamin D3, have a relatively high risk of hypercalcaemia, and have a relatively

Sufficient 25(OH)D 30–50 ng/mL or 75–125 nmol/L
• Moderate sunlight exposure is recommended, with adequate dietary intake of vitamin D. The use of vitamin D preparations is recommended depending on the season, BMI, age, skin colour, and risk factors.
• During the winter months from November to April, the use of 800–2000 IU/day of vitamin D is recommended in people without risk factors
• Administration throughout the year of 800–2000 IU/day of vitamin D is recommended in: • women planning pregnancy, during prolonged hospitalizations, in patients over 65 years of age, in patients with osteoporosis or the risk of falling and fractures, as maintenance dose after treatment of vitamin D deficiency
• 2–3 times higher doses of vitamin D are recommended in people with malabsorption syndrome, obese (BMI >30 kg/m2), and persons with black skin colour.

Table 1.
Preventive use of vitamin D.

Insufficient $\geq 20 - < 30\ ng/ml$ ($\geq$50-<75 nmol/L) or deficient <20 ng/mL(<50 nmol/L)
• If clinically indicated rapid 25(OH)D correction in patients with vitamin D deficiency without significant risk factors, it is recommended to use 6000 IU D3/day. In certain persons, high doses of vitamin D-10,000 IU D3/day are applied: patients with malabsorption syndrome, patients using chronic therapy with drugs that disrupt vitamin D metabolism and obese persons (BMI > 30 kg/m^2)
• The treatment lasts 4–12 weeks depending on the degree of vitamin D deficiency.
• If the recommended vitamin D level is reached, 30–50 ng/mL or 75–125 nmol/L, the administration of 800–2000 UI/day of the vitamin D is continued.
• Approximately 6–12 weeks after starting treatment, the level of 25(OH)D is checked.

Table 2.
Therapeutic doses of vitamin D.

narrow therapeutic window. Active treatment with calcitriol is indicated only in certain diseases, for example, chronic hypoparathyroidism and chronic kidney disease–mineral and bone disorder (CKD-MBD) [67, 72, 73].

4.1 Side effects of vitamin D

Vitamin D intoxication occurs at a level of 25(OH)D > 375 nmol/l (150 ng/ml) and is a very rare complication. The first manifestation of vitamin D intoxication is calciuria, caused by reducing the reabsorption of calcium in the proximal tubules. When the compensatory mechanisms can no longer increase calciuria, its concentration in the blood increases. Hypercalcaemia decreases the parathyroid hormone (PTH) level, which consequently reduces renal phosphorus excretion. A high concentration of 25 (OH)D directly through VDR in the intestine further increases intestinal absorption of calcium and phosphorus, which further increases their blood level, and as a result calcium/phosphorus solubility product increases, leading to calcifications in soft tissues and renal tissue (nephrocalcinosis), and also in the blood vessels. Hypercalcaemia also causes constipation, vasoconstriction with subsequent arterial hypertension, cardiac arrhythmias, depression, confusion, polyuria, and polydipsia [35, 74].

Therefore, in patients with advanced DN and significantly reduced GFR, care should be taken when supplementing vitamin D, that is, monitor the level of calcium and phosphorus in the blood. If the phosphorus level is elevated, stop the administration of vitamin D, until the phosphorus in the blood normalizes and include one of the phosphorus-binding preparations in the therapy. Also in case of hypercalcemia, discontinue the administration of vitamin D.

5. Conclusion

We can conclude that the use of vitamin D in patients with DM and DN has a significant renoprotective effect, both in patients with albuminuria and in patients without albuminuria. The particularly significant renoprotective effect of vitamin D is reflected in the stabilization of podocyte function. Vitamin D has also cardiovascular protection by decreasing albuminuria, correcting lipid status, and improving parameters of glycoregulation and inflammation.

In order to obtain more data on the renoprotective effect of vitamin D, additional studies are needed, including more patients and longer follow-up.

Conflict of interest

The authors declare no conflict of interest.

Acronyms and abbreviations

1,25(OH)2D3	1,25- hydroxyvitamin D3
25(OH)D3	25-hydroxyvitamin D3
ACEi	Angiotensin-converting enzyme inhibitors
ADA	American Diabetes Association
AGEs	Advanced glycation end products

AT1R	Angiotensin II type 1 receptor
BMI	Body mass index
CIP27B1	1-alpha-hydroxylase
CKD-MBD	Chronic kidney disease–mineral and bone disorder
CRE	Cyclic adenosine monophosphate-response element
DM	Diabetes mellitus
DN	Diabetic nephropathy
DR	Diabetic retinopathy
ESRD	End-stage renal disease
GFR	Glomerular filtration rate
HbA1c	Glycated haemoglobin A1c
ICAM-1	Intracellular adhesion molecule 1
IDF	International Diabetes Federation
IGF-1	Insulin-like growth factor 1
IGT	Impaired Glucose Tolerance
INF-γ	Interferon-gamma
IL	Interleukin
KIM-1	Kidney injury molecule-1
L-FABP	Liver-type fatty acid binding protein
LDL	Low-density lipoprotein
LMICs	Low- and middle-income countries
M	Macrophages
MCP-1	Monocyte chemoattractant protein-1
MMP-2	Matrix metalloproteinase-2
$Na+/K+$-ATP-ase	Sodium potassium adenosine triphosphatase
NAD+	Nicotinamide dinucleotide
NADPH	Nicotinamide dinucleotide phosphate
NGAL	Neutrophil gelatinase-associated lipocalin
NKF KDOQI	National Kidney Foundation Kidney Disease Outcomes Quality Initiative
NP-DN	Non-proteinuric diabetic nephropathy
P-DN	Proteinuric diabetic nephropathy
PDGF	Platelet-derived growth factor
PICK1	Protein interacting with C kinase 1
PKC	Protein kinase C
PPAR δ	Peroxisome proliferator activator receptor δ
PTH	Parathyroid hormone
RAAS	Renin-angiotensin-aldosterone system
RANTES	Regulated upon activation, Normal T cell expressed and presumably Secreted
ROS	Reactive oxygen species
SD	Slit diaphragm
SGLT1	Sodium-glucose cotransporter-1
SGLT2	Sodium-glucose cotransporter-2
T1DM	Type 1 diabetes mellitus
T2DM	Type 2 diabetes mellitus
TGF	Tubuloglomerular feedback
TGF-β1	Transforming growth factor Beta 1
Th	T lymphocytes
TNF-α	Tumor necrosis factor-alpha

UACR	Urine albumin-to-creatinine ratio
VCAM-1	Vascular cell adhesion molecule-1
VDBP	Vitamin D binding protein
VEGF	Vascular endothelial growth factor

Author details

Marijana Petrovic*, Stanko Petrovic and Katarina Obrencevic
Military Medical Academy, Belgrade, Serbia

*Address all correspondence to: makystan@yahoo.com

References

[1] Khan MAB, Hashim MJ, King JK, Govender RD, Mustafa H, Al KJ. Epidemiology of Type 2 diabetes—Global burden of disease and forecasted trends. Journal of Epidemiology and Global Health. 2020;**10**(1):107-111. DOI: 10.2991/jegh.k.191028.001

[2] Sufra R, Lookens Pierre J, Dade E, Rouzier V, Apollon A, St. Preux S, et al. Diabetes epidemiology among adults in Port-au-Prince, Haiti: A Cross-Sectional Study. Frontiers in Endocrinology. 2022; **13**:841675

[3] International Diabetes Federation. IDF Diabetes Atlas, 10th Edn. [Internet]. 2021. Brussels, Belgium. Available from: https://www.diabetesatlas.org [Accessed: November 15, 2021]

[4] Saeedi P, Petersohn I, Salpea P, Malanda B, Karuranga S, Unwin N, et al. IDF Diabetes Atlas Committee. Global and regional diabetes prevalence estimates for 2019 and projections for 2030 and 2045: Results from the International Diabetes Federation Diabetes Atlas, 9th edition. Diabetes Researc Clinical Practise. 2019;**157**: 107843

[5] NCD Risk Factor Collaboration (NCD-RisC). Worldwide trends in diabetes since 1980: A pooled analysis of 751 population-based studies with 4.4 million participants. Lancet. 2016; **387**(10027):1513-1530

[6] Wu Y, Ding Y, Tanaka Y, Zhang W. Risk factors contributing to type 2 diabetes and recent advances in the treatment and prevention. International Journal of Medical Sciences. 2014;**11**(11): 1185-1200. DOI: 10.7150/ijms.10001

[7] Traversi D, Rabbone I, Scaioli G, Vallini C, Carletto G, Racca I, et al. Risk factors for type 1 diabetes, including environmental, behavioural and gut microbial factors: A case-control study. Scientific Reports. 2020;**10**(1):17566. DOI: 10.1038/s41598-020-74678-6

[8] Chawla A, Chawla R, Jaggi S. Microvasular and macrovascular complications in diabetes mellitus: Distinct or continuum? Indian Journal of Endocrinology and Metabolism. 2016; **20**(4):546

[9] United States Renal Data System. 2016 USRDS Annual Data Report: Epidemiology of Kidney Disease in the United States. Bethesda, MD: National Institutes of Health, National Institute of Diabetes and Digestive and Kidney Diseases; 2016

[10] ERA-EDTA Registry. ERA-EDTA Registry Annual Report 2015. Amsterdam, the Netherlands, Academic Medical Center, Department of Medical Informatics; 2017

[11] Lim AK. Diabetic nephropathy—Complications and treatment. International Journal of Nephrology and Renovascular Disease. 2014;**7**:361-381. DOI: 10.2147/IJNRD.S40172

[12] Yamanouchi M, Furuichi K, Hoshino J, Ubara Y, Wada T. Nonproteinuric diabetic kidney disease. Clinical and Experimental Nephrology. 2020;**24**(7):573-581. DOI: 10.1007/s10157-020-01881-0

[13] Yamanouchi M, Furuichi K, Hoshino J, Toyama T, Hara A, Shimizu M, et al. Nonproteinuric versus proteinuric phenotypes in diabetic kidney disease: A Propensity Score-Matched Analysis of a Nationwide, Biopsy-Based Cohort Study. Diabetes

Care. 2019;**42**(5):891-902. DOI: 10.2337/dc18-1320

[14] Lin YC, Chang YH, Yang SY, Wu KD, Chu TS. Update of pathophysiology and management of diabetic kidney disease. Journal of the Formosan Medical Association. 2018;**117**(8):662-675. DOI: 10.1016/j.jfma.2018.02.007

[15] Novak JE, Yee J. Diabetes and the kidney: Sweet dreams. Advances in Chronic Kidney Disease. 2018;**25**(2):115-118. DOI: 10.1053/j.ackd.2018.01.001

[16] Fioretto P, Caramori ML, Mauer M. The kidney in diabetes: Dynamic pathways of injury and repair. The Camillo Golgi Lecture 2007. Diabetologia. 2008;**51**(8):1347-1355. DOI: 10.1007/s00125-008-1051-7

[17] Carranza K, Veron D, Cercado A, Bautista N, Pozo W, Tufro A, et al. Cellular and molecular aspects of diabetic nephropathy; the role of VEGF-A. Nefrologia. 2015;**35**(2):131-138. DOI: 10.1016/j.nefro.2015.05.013

[18] Galicia-Garcia U, Benito-Vicente A, Jebari S, Larrea-Sebal A, Siddiqi H, Uribe KB, et al. Pathophysiology of Type 2 diabetes mellitus. International Journal of Molecular Sciences. 2020;**21**(17):6275. DOI: 10.3390/ijms21176275

[19] Teng B, Duong M, Tossidou I, Yu X, Schiffer M. Role of protein kinase C in podocytes and development of glomerular damage in diabetic nephropathy. Frontiers in Endocrinology. 2014;**5**:179

[20] Samsu N. Diabetic nephropathy: Challenges in pathogenesis, diagnosis, and treatment. BioMed Research International. 2021;**2021**:1497449. DOI: 10.1155/2021/1497449

[21] Quack I, Woznowski M, Potthoff SA, Palmer R, Königshausen E, Sivritas S, et al. PKC alpha mediates beta-arrestin2-dependent nephrin endocytosis in hyperglycemia. The Journal of Biological Chemistry. 2011;**286**(15):12959-12970. DOI: 10.1074/jbc.M110.204024

[22] Anil Kumar P, Welsh GI, Saleem MA, Menon RK. Molecular and cellular events mediating glomerular podocyte dysfunction and depletion in diabetes mellitus. Frontiers in Endocrinology (Lausanne). 2014;**5**:151

[23] Duran-Salgado MB, Rubio-Guerra AF. Diabetic nephropathy and inflammation. World Journal of Diabetes. 2014;**5**(3):393-398. DOI: 10.4239/wjd.v5.i3.393

[24] Marrero MB, Banes-Berceli AK, Stern DM, Eaton DC. Role of the JAK/STAT signaling pathway in diabetic nephropathy. American Journal of Physiology. Renal Physiology. 2006;**290**(4):F762-F768. DOI: 10.1152/ajprenal.00181.2005

[25] National Kidney Foundation. KDOQI Clinical Practice Guideline for Diabetes and CKD: 2012 Update. American Journal of Kidney Diseases. 2012;**60**(5):850-856

[26] Araki S, Haneda M, Sugimoto T, Isono M, Isshiki K, Kashiwagi A, et al. Factors associated with frequent remission of microalbuminuria in patients with type 2 diabetes. Diabetes. 2005;**54**(10):2983-2987. DOI: 10.2337/diabetes.54.10.2983

[27] Gaede P, Tarnow L, Vedel P, Parving HH, Pedersen O. Remission to normoalbuminuria during multifactorial treatment preserves kidney function in patients with type 2 diabetes and microalbuminuria. Nephrology, Dialysis,

Transplantation. 2004;**19**(11):2784-2788. DOI: 10.1093/ndt/gfh470

[28] Chang DY, Li MR, Yu XJ, Wang SX, Chen M, Zhao MH. Clinical and pathological characteristics of patients with nonproteinuric diabetic nephropathy. Frontiers in Endocrinology. 2021;**12**:761386

[29] Dwyer JP, Parving HH, Hunsicker LG, Ravid M, Remuzzi G, Lewis JB. Renal dysfunction in the presence of normoalbuminuria in Type 2 Diabetes: Results from the DEMAND Study. Cardiorenal Medical. 2012;**2**(1): 1-10. DOI: 10.1159/000333249

[30] Krolewski AS, Niewczas MA, Skupien J, Gohda T, Smiles A, Eckfeldt JH, et al. Early progressive renal decline precedes the onset of microalbuminuria and its progression to macroalbuminuria. Diabetes Care. 2014; **37**(1):226-234. DOI: 10.2337/dc13-0985

[31] Reutens AT. Epidemiology of diabetic kidney disease. The Medical Clinics of North America. 2013;**97**(1): 1-18. DOI: 10.1016/j.mcna.2012.10.001

[32] Romano G, Mioni R, Danieli N, Bertoni M, Croatto E, Merla L, et al. Elevated intrarenal resistive index predicted faster renal function decline and long-term mortality in non-proteinuric chronic kidney disease. Journal of Clinical Medicine. 2022; **11**(11):2995. DOI: 10.3390/jcm11112995

[33] Kopel J, Pena-Hernandez C, Nugent K. Evolving spectrum of diabetic nephropathy. World Journal of Diabetes. 2019;**10**(5):269-279. DOI: 10.4239/ wjd.v10.i5.269

[34] Tervaert TW, Mooyaart AL, Amann K, Cohen AH, Cook HT, Drachenberg CB, et al. Pathologic classification of diabetic nephropathy. Journal of American Society of Nephrology. 2010;**21**(4):556-563. DOI: 10.1681/ASN.2010010010

[35] Pludowski P, Holick MF, Grant WB, Konstantynowicz J, Mascarenhas MR, Haq A, et al. Vitamin D supplementation guidelines. The Journal of Steroid Biochemistry and Molecular Biology. 2018;**175**:125-135. DOI: 10.1016/j. jsbmb.2017.01.021

[36] Christakos S, Dhawan P, Verstuyf A, Verlinden L, Carmeliet G. Vitamin D: Metabolism, molecular mechanism of action, and pleiotropic effects. Physiological Reviews. 2016;**96**(1): 365-408. DOI: 10.1152/ physrev.00014.2015

[37] Bover J, Egido J, Fernández-Giráldez E, Praga M, Solozábal-Campos C, Torregrosa JV, et al. Vitamin D, vitamin D receptor and the importance of its activation in patients with chronic kidney disease. Nefrologia. 2015;**35**(1): 28-41. DOI: 10.3265/Nefrologia.pre2014. Sep.11796

[38] Delrue C, Speeckaert R, Delanghe JR, Speeckaert MM. The role of Vitamin D in diabetic nephropathy: A Translational Approach. International Journal of Molecular Sciences. 2022; **23**(2):807. DOI: 10.3390/ijms23020807

[39] Peng Y, Li LJ. Serum 25-hydroxyvitamin D level and diabetic nephropathy in patients with type 2 diabetes mellitus. International Urology and Nephrology. 2015;**47**(6):983-989. DOI: 10.1007/s11255-015-0983-3

[40] Petrović M, Dragović T, Petrović S, Obrenčević K, Rančić N, Đurašinović T, et al. Effect of vitamin d on proteinuria, lipid status, glicoregulation and c-reactive protein in patients with type-2 diabetes mellitus. VojnosanitPregl. 2020; **77**(6):582-589

[41] Zhang Z, Sun L, Wang Y, Ning G, Minto AW, Kong J, et al. Renoprotective role of the vitamin D receptor in diabetic nephropathy. Kidney International. 2008;**73**(2):163-171. DOI: 10.1038/sj.ki.5002572

[42] Li X, Liu Y, Zheng Y, Wang P, Zhang Y. The effect of Vitamin D supplementation on glycemic control in Type 2 diabetes patients: A systematic review and meta-analysis. Nutrients. 2018;**10**(3):375. DOI: 10.3390/nu10030375

[43] von Hurst PR, Stonehouse W, Coad J. Vitamin D supplementation reduces insulin resistance in South Asian women living in New Zealand who are insulin resistant and vitamin D deficient—A randomised, placebo-controlled trial. The British Journal of Nutrition. 2010;**103**(4): 549-555. DOI: 10.1017/S0007114509992017

[44] Park SK, Garland CF, Gorham ED, BuDoff L, Barrett-Connor E. Plasma 25-hydroxyvitamin D concentration and risk of type 2 diabetes and pre-diabetes: 12-year cohort study. PLoS One. 2018; **13**(4):e0193070. DOI: 10.1371/journal.pone.0193070

[45] Afzal S, Bojesen SE, Nordestgaard BG. Low 25-hydroxyvitamin D and risk of type 2 diabetes: A prospective cohort study and metaanalysis. Clinical Chemistry. 2013; **59**(2):381-391. DOI: 10.1373/clinchem.2012.193003

[46] Tabesh M, Azadbakht L, Faghihimani E, Tabesh M, Esmaillzadeh A. Effects of calcium-vitamin D co-supplementation on metabolic profiles in vitamin D insufficient people with type 2 diabetes: A randomised controlled clinical trial. Diabetologia. 2014;**57**(10):2038-2047. DOI: 10.1007/s00125-014-3313-x

[47] Garbossa SG, Folli F. Vitamin D, sub-inflammation and insulin resistance. A window on a potential role for the interaction between bone and glucose metabolism. Reviews in Endocrine & Metabolic Disorders. 2017;**18**(2): 243-258. DOI: 10.1007/s11154-017-9423-2

[48] Humalda JK, Goldsmith DJ, Thadhani R, de Borst MH. Vitamin D analogues to target residual proteinuria: Potential impact on cardiorenal outcomes. Nephrology, Dialysis, Transplantation. 2015;**30**(12):1988-1994. DOI: 10.1093/ndt/gfu404

[49] Momeni A, Mirhosseini M, Kabiri M, Kheiri S. Effect of vitamin D on proteinuria in type 2 diabetic patients. Journal of Nephropathology. 2017;**6**(1): 10-14

[50] de Zeeuw D, Agarwal R, Amdahl M, Audhya P, Coyne D, Garimella T, et al. Selective vitamin D receptor activation with paricalcitol for reduction of albuminuria in patients with type 2 diabetes (VITAL study): A randomised controlled trial. Lancet. 2010;**376**(9752): 1543-1551. DOI: 10.1016/S0140-6736(10)61032-X

[51] Schmitz KJ, Skinner HG, Bautista LE, Fingerlin TE, Langefeld CD, Hicks PJ, et al. Association of 25-hydroxyvitamin D with blood pressure in predominantly 25-hydroxyvitamin D deficient Hispanic and African Americans. American Journal of Hypertension. 2009;**22**(8):867-870. DOI: 10.1038/ajh.2009.88

[52] Gembillo G, Cernaro V, Salvo A, Siligato R, Laudani A, Buemi M, et al. Role of Vitamin D status in diabetic patients with renal disease. Medicina (Kaunas, Lithuania). 2019;**55**(6):273. DOI: 10.3390/medicina55060273

[53] Ramiro-Lozano JM, Calvo-Romero JM. Effects on lipid profile of supplementation with vitamin D in type 2 diabetic patients with vitamin D deficiency. Therapeutic Advances in Endocrinology and Metabolism. 2015; **6**(6):245-248. DOI: 10.1177/2042018815599874

[54] Qin XF, Zhao LS, Chen WR, Yin DW, Wang H. Effects of vitamin D on plasma lipid profiles in statin-treated patients with hypercholesterolemia: A randomized placebo-controlled trial. Clinical Nutrition. 2015;**34**(2):201-206. DOI: 10.1016/j.clnu.2014.04.017

[55] Muñoz-Aguirre P, Flores M, Macias N, Quezada AD, Denova-Gutiérrez E, Salmerón J. The effect of vitamin D supplementation on serum lipids in postmenopausal women with diabetes: A randomized controlled trial. Clinical Nutrition. 2015;**34**(5):799-804. DOI: 10.1016/j.clnu.2014.10.002

[56] Sollid ST, Hutchinson MY, Fuskevåg OM, Figenschau Y, Joakimsen RM, Schirmer H, et al. No effect of high-dose vitamin D supplementation on glycemic status or cardiovascular risk factors in subjects with prediabetes. Diabetes Care. 2014; **37**(8):2123-2131. DOI: 10.2337/dc14-0218

[57] Chang AS, Hathaway CK, Smithies O, Kakoki M. Transforming growth factor-β1 and diabetic nephropathy. American Journal of Physiology. Renal Physiology. 2016; **310**(8):F689-F696. DOI: 10.1152/ajprenal.00502.2015

[58] Di Paolo S, Gesualdo L, Ranieri E, Grandaliano G, Schena FP. High glucose concentration induces the overexpression of transforming growth factor-beta through the activation of a platelet-derived growth factor loop in human mesangial cells. The American Journal of Pathology. 1996;**149**(6): 2095-2106

[59] Kim MJ, Frankel AH, Donaldson M, Darch SJ, Pusey CD, Hill PD, et al. Oral cholecalciferol decreases albuminuria and urinary TGF-β1 in patients with type 2 diabetic nephropathy on established renin-angiotensin-aldosterone system inhibition. Kidney International. 2011; **80**(8):851-860. DOI: 10.1038/ki.2011.224

[60] Tian Y, Lv G, Yang Y, Zhang Y, Yu R, Zhu J, et al. Effects of vitamin D on renal fibrosis in diabetic nephropathy model rats. International Journal of Clinical and Experimental Pathology. 2014;7(6):3028-3037

[61] Wang R, Yao C, Liu F. Association between renal podocalyxin expression and renal dysfunction in patients with diabetic nephropathy: A Single-Center, Retrospective Case-Control Study. Biomedical Research International. 2020; **2020**:7350781. DOI: 10.1155/2020/7350781

[62] Petrica L, Vlad A, Gluhovschi G, Gadalean F, Dumitrascu V, Gluhovschi C, et al. Proximal tubule dysfunction is associated with podocyte damage biomarkers nephrin and vascular endothelial growth factor in type 2 diabetes mellitus patients: A cross-sectional study. PLoS One. 2014;**9**(11): e112538. DOI: 10.1371/journal.pone.0112538

[63] Uchida K, Suzuki K, Iwamoto M, Kawachi H, Ohno M, Horita S, et al. Decreased tyrosine phosphorylation of nephrin in rat and human nephrosis. Kidney International. 2008;**73**(8): 926-932. DOI: 10.1038/ki.2008.19

[64] Tufro A, Veron D. VEGF and podocytes in diabetic nephropathy. Seminars in Nephrology. 2012;**32**(4):

385-393. DOI: 10.1016/j.semnephrol.2012.06.010

[65] Gil CL, Hooker E, Larrivée B. Diabetic kidney disease, endothelial damage, and podocyte-endothelial crosstalk. Kidney Medical. 2020;**3**(1): 105-115. DOI: 10.1016/j.xkme.2020.10.005

[66] Seida JC, Mitri J, Colmers IN, Majumdar SR, Davidson MB, Edwards AL, et al. Clinical review: Effect of vitamin D3 supplementation on improving glucose homeostasis and preventing diabetes: A systematic review and meta-analysis. The Journal of Clinical Endocrinology and Metabolism. 2014;**99**(10):3551-3560. DOI: 10.1210/jc.2014-2136

[67] Pludowski P, Takacs I, Boyanov M, Belaya Z, Diaconu CC, Mokhort T, et al. Clinical practice in the prevention, diagnosis and treatment of Vitamin D deficiency: A Central and Eastern European Expert Consensus Statement. Nutrients. 2022;**14**(7):1483. DOI: 10.3390/nu14071483

[68] Rusińska A, Płudowski P, Walczak M, Borszewska-Kornacka MK, Bossowski A, Chlebna-Sokół D, et al. Vitamin D Supplementation Guidelines for General Population and Groups at Risk of Vitamin D Deficiency in Poland-Recommendations of the Polish Society of Pediatric Endocrinology and Diabetes and the Expert Panel With Participation of National Specialist Consultants and Representatives of Scientific Societies-2018 Update. Frontiers in Endocrinology. 2018;**9**:246

[69] Haq A, Wimalawansa SJ, Pludowski P, Anouti FA. Clinical practice guidelines for vitamin D in the United Arab Emirates. The Journal of Steroid Biochemistry and Molecular Biology. 2018;**175**:4-11. DOI: 10.1016/j.jsbmb.2016.09.021

[70] Sosa Henríquez M, Gómez de Tejada Romero MJ. Cholecalciferol or calcifediol in the management of Vitamin D deficiency. Nutrients. 2020;**13**:1617

[71] Pérez-Castrillón JL, Dueñas-Laita A, Brandi ML, Jódar E, Del Pino-Montes J, Quesada-Gómez JM, et al. Calcifediol is superior to cholecalciferol in improving vitamin D status in postmenopausal women: A randomized trial. Journal of Bone and Mineral Research. 2021;**36**(10):1967-1978. DOI: 10.1002/jbmr.4387

[72] Bollerslev J, Rejnmark L, Zahn A, Heck A, Appelman-Dijkstra NM, Cardoso L, et al. European Expert Consensus on Practical Management of Specific Aspects of Parathyroid Disorders in Adults and in Pregnancy: Recommendations of the ESE Educational Program of Parathyroid Disorders. European Journal of Endocrinology. 2022;**186**(2):R33-R63. DOI: 10.1530/EJE-21-1044

[73] Ramasamy I, Vitamin D. Metabolism and guidelines for Vitamin D supplementation. Clinical Biochemical Review. 2020;**41**(3):103-126

[74] Holick MF, Vitamin D. Is not as toxic as was once thought: A historical and an up-to-date perspective. Mayo Clinic Proceedings. 2015;**90**(5):561-564. DOI: 10.1016/j.mayocp.2015.03.015

Chapter 10

Vitamin D, Muscle Strength and Cardiorespiratory Fitness – An Evidence-based Review

Amin Mirrafiei, Mahsa Firouzi, Nadia Babaee, Samira Davarzani and Sakineh Shab-Bidar

Abstract

Recent evidence reported that a higher concentration of 25-hydroxyvitamin D [25[OH] D] has been associated with greater cardiorespiratory fitness [CRF] and muscle strength in both sexes. Low levels of 25[OH]D may be related to hypertrophy of myocardial, high blood pressure, and endothelial dysfunction, which is related to decreased amino acid uptake, prolonged time to peak muscle contraction and relaxation, dysregulation of intracellular Ca2+, muscle weakness, myalgia, impaired neuromuscular function, and hypotonia. Because CRF is defined as a function of maximal cardiac output and maximal arteriovenous oxygen difference, low levels of 25[OH]D may lead to deleterious effects on CRF. Recent findings also indicated vitamin D3 supplementation that leads to an increase in muscle fiber especially type 2, the cross-sectional area of muscle fibers, and improved muscle strength. In this chapter, we will systematically review the observational studies and randomized controlled trials that evaluated the association of vitamin D with CRF and muscle strength.

Keywords: vitamin D, 25-hydroxyvitamin D, Vo2 max, fitness, muscle strength

1. Introduction

Vitamin D deficiency has become a major public health issue around the world affecting approximately 1 billion individuals worldwide. Vitamin D has long been known to play a role in the performance of many organs and tissues throughout the human body [1]. A low serum vitamin D level has been linked to an increased risk of diabetes, hypertension, and cardiovascular disease [CVD], as well as obesity, hyperlipidemia, and poor physical fitness, particularly low cardiorespiratory endurance and muscle strength [2, 3]. It has been documented in medical guidelines that vitamin D enhances muscular function in poor vitamin D conditions [4].

Cardiorespiratory fitness [CRF] is a physiologic fitness definition that refers to the circulatory and respiratory systems' ability to deliver oxygen during prolonged physical activity. It is partially determined by several non-modifiable characteristics such as gender, age, and hereditary factors. CRF also has been used to analyze the association between physical activity and health status in recent years as a mediator [5]. In addition, muscle strength is a marker of the quality of functional performance. A decline in muscle strength leads to physical malfunction and disability in parallel to age and is associated with mobility restriction [6]. It has been debated that greater muscular strength may improve exercise performance by allowing for higher levels of cardiorespiratory stress [7].

A higher quantity of 25-hydroxyvitamin D [25[OH] D] has been linked to increased CRF and muscle strength in both sexes, according to accumulative research. Low amounts of 25[OH]D may have a negative influence on CRF. Furthermore, previous research has found a substantial link between serum vitamin D levels and physical fitness [8, 9]. The current chapter discusses the most recent knowledge about the link between vitamin D status, CRF, and muscle strength.

2. What is vitamin D?

Vitamin D is a fat-soluble component that can be synthesized via sun exposure to the skin. The ultraviolet rays in sunlight make your skin make vitamin D [10]. Of course, the amount of production of this vitamin by your skin depends on various factors. For instance, what season of the year and what time of the day you are exposed to sunlight, is effective in the production of this vitamin by your skin. Note that usually, the sun's rays are less in the winter months. Also, the sun's rays are strongest between 10 am and 3 pm. The amount of cloud cover and air pollution and geographical location also have a significant effect on the production of vitamin D in the skin, as sites near the equator have higher levels of UV radiation. Vitamin D production in the skin is reduced by more than 95% when using sunscreen. To produce the same quantity of vitamin D as someone with white skin, someone with a naturally dark skin tone needs to be exposed to the sun for at least three to five times longer [11–14]. Furthermore, obesity is linked to vitamin D deficiency since there is an inverse relationship between serum 25[OH]D, activated in the liver, and a body mass index [BMI] of more than 30 kg/m^2 [15]. The synthesized vitamin D in human skin is called D3, which is also found in salmon, sardine, mackerel, tuna, liver, egg yolk, and fortified foods like milk [16]. Another type of vitamin D, dubbed D2, is abundant in mushrooms [17]. The majority of the body's tissues and cells include the vitamin D receptor [VDR]. Numerous biological processes are influenced by 1,25[OH]2D, the active form of vitamin D that is transformed in the kidney, including the inhibition of cellular proliferation, induction of terminal differentiation, inhibition of angiogenesis, stimulation of insulin production, and inhibition of renin production [18]. Up to 200 genes that are thought to be involved in many of the health-related functions of vitamin D may be controlled by the local synthesis of 1,25[OH]2D [18]. The Institute of Medicine [IOM] recommends that vitamin D deficiency is defined as serum 25[OH]D concentration < 50 nmol/L and vitamin D sufficiency as 50 nmol/L, and optimal level is >75 nmol/L [1].

2.1 Relationship with cardiorespiratory fitness

Besides many biochemical and physiological properties of active vitamin D in the body, this vitamin can have huge effects on CRF, as recent research revealed a significant association between serum 25[OH]D levels and CRF. In a recent systematic review and meta-analysis [To find the answers to a particular topic, a systematic review makes an effort to compile all accessible empirical studies. The statistical method of assessing and combining data from numerous related studies is called a meta-analysis [19].], our team that included both observational and interventional studies [up to October 2018, **Tables 1** and **2**] showed that in observational studies, serum 25[OH]D is directly related to CRF, as shown in **Figure 1,** with a significant increase in CRF in line with an increased 25[OH]D level [+0.65]. Furthermore, it was found through reanalysis of five clinical trials that vitamin D3 treatment raised CRF in comparison to placebo (**Figure 2**).

VO2 max, one of the most often used tests to quantify endurance capacity, is used to assess cardiorespiratory fitness. VO2 max stands for the maximal capacity to transport and utilize oxygen during exercise performed at increasing intensities. In other words, the maximum rate of oxygen consumption that may be achieved during intense activity is known as VO2 max [20]. It is used to describe the intensity of the aerobic process and displays the level of physical preparedness of an athlete. VO2 max is typically evaluated in laboratories on treadmills, cycling ergometers, or rowing ergometers by gradually increasing intensity over some time of more than 5 minutes [21]. Cardiac output, arterial oxygen content, the blood supply to active muscles, and oxygen use by muscles all contribute to VO2 max [3]. Through the action of vitamin D receptors, low serum 25[OH]D levels can result in cardiac hypertrophy, increased blood pressure, and endothelial dysfunction. It can, therefore, affect VO2 max by lowering cardiac output and raising peripheral vascular resistance [22]. Exercise raises VO2 max by boosting cardiac output. Those with modest levels of physical exercise may benefit more from vitamin D in terms of cardiac remodeling and VO2 max [23]. Furthermore, vitamin D insufficiency and physical inactivity can promote muscular atrophy and change the muscle fiber type [24].

The mechanisms behind Vitamin D's beneficial effects are the increased numbers of fast-twitch muscle fibers [IIa] in place of another type of fast-twitch muscle fibers [IIb], modification of maximum heart rate and stroke volume. In addition, because low levels of 25[OH]D affect bone mineralization and muscle function, they are linked to a decline in physical fitness. Vitamin D may have a function in lowering cortisol by preventing specific enzymes from working. High levels of cortisol boost anti-inflammatory and calcification effects. By causing the dilatation of blood vessels, a decreased cortisol level frequently worsens blood pressure. By lowering blood cortisol levels, vitamin D aids in enhancing physical performance and lowering cardiovascular risk factors. Additionally, inflammation is decreased and interleukin-10 is produced by vitamin D. Thus, the probable mechanisms that define the effect of this vitamin on maximum oxygen intake are expanded airways, antimicrobial peptides, and greater air entry into the lungs by vitamin D [25–27].

2.2 Relationship with muscle strength

Skeletal muscle helps organ systems maintain homeostasis. Muscle is malleable, adapting to physical activity, load, injury, sickness, and aging. The reduction of skeletal muscular strength, muscle mass, and physical performance as people get older has been linked to falls and fractures in elderly people, yet it is still a generally undetected disorder [28].

The presence of vitamin D3 metabolizing enzymes in skeletal muscle raises the possibility that vitamin D3 levels are locally regulated in this extrarenal tissue [28]. Total fat mass, lean mass, and balance are all physical fitness indices that are commonly influenced by vitamin D levels. Studies have demonstrated that severe

Study first author	Country	Year	Journal	Study population	Sex	Sample size	Range age, or mean age	Assess vitamin D	Assess vo2 max	Correlation r	Adjusted
Al Asoom	KSA	2016	Journal of Taibah University Medical Sciences	Young Saudi females	Female	87	20.78	Liquid chromatography (HPLC)	Bruce treadmill protocol	0.259	Adjusted but Not reported
Mowry	USA	2009	J Am Osteopath Assoc	Adolescent girls and young women	Female	59	16_24	Chemiluminescent	Maximal-graded treadmill test	0.36	Not reported
Chang-Duk	Korea	2013	Med Sci Sports Exerce	Young and healthy college male student	Male	799	24.2 24.2 23.7	LIAISON 25(OH) vitamin D total assay (CLIA)	Maximal graded exercise test on a treadmill	0.4	Un adjusted
Valtuen˜a	Europe	2013	Qjm	Random sample of 3000 European adolescents	Male	470	12.5–17.5	ELISA	20 m shuttle run test	0.108	Not reported
Valtuen˜a	Europe	2013	Qjm	Random sample of 3000 European adolescents	Female	536	12.5–17.5	ELISA	20 m shuttle run test	0.022	Not reported
Ellis	USA	2014	Endocrine	Healthy women	Female	67	60–74	Immunoassay	Modified bruce graded treadmill protocol	0.344	Partial adjusted for percent fat
Serra	USA	2016	Hormone and Metabolic Research	Overweight and obese sedentary women with history of GDM	Female	51	38–65	By RIA (DiaSorin, Still water MN)	Graded exercise test on a treadmill	0.26	Not reported
Koundourakis	Greece	2014	PLoS One	Members of two teams and one Football team	Male	67	25.6	DiaSorin25 hydroxy vitamin D	Motorizetreadmill using an automated gas-analysis system	0.436	Not reported

Study first author	Country	Year	Journal	Study population	Sex	Sample size	Range age, or mean age	Assess vitamin D	Assess vo2 max	Correlation r	Adjusted
Dong	USA	2010	J Exerc Nutrition Biochem	Adolescents from high schools,	Male and female	559	14–18	Liquid chromatography–mass spectroscopy	Multi stage treadmill test	0.1	For age, sex, race, sexual maturation, height, and season
Waschbisch	Switzerland	2012	Eur Neurol	Patients with relapsing–remitting disease	Male and female	42	39 36	ELISA	Bicycle ergometer	0.4	Not reported
Park	Korea	2013	J Exerc Nutrition Biochem	Male students in the university of S in Jangan-gu, Suwon, Gyeonggi	Male	593	24.2 24.2 23.7	DiaSorin LIAISON automated analyzer	Bruce treadmill protocol	0.326	Not reported
Sun	Japan	2014	Nutrients	One hundred and seven Japanese men	Male	107	40–79	Immunosorbent assay	Graded exercise test on cycle ergometer	0.34	Age, season, VFA
Sun	Japan	2015	J Atheroscler Thromb	136 healthy Japanese men	Male	136	20–79	Semi-automated device	Cycle ergometer	0.361	Age adjusted
Ardestani	USA	2011	Am J Cardiol	Adults free of overt cardiovascular and metabolic disease	Male and female	200	40	ELISA immunoassay protocol	Modified Balke treadmill test	0.29	Not reported
. Fitzgerald	USA	2014	Journal of Strength and Conditioning Research	Fifty-seven Caucasian male competitive ice hockey players	Male	52	18–23	Liquid chromatography-spectrometry	During a skate treadmill GXT.	0.052	

Study first author	Country	Year	Journal	Study population	Sex	Sample size	Range age, or mean age	Assess vitamin D	Assess vo2 max	Correlation r	Adjusted
Saponaro	Italy	2017	Endocrine	Consecutive patients diagnosed with heart failure	Male and female	261	65	HPLC-MS/MS	Electronically braked cycle-ergometer by a ramp protocol	0.16	Un adjusted
Marawan	USA	2018	European Journal of Preventive Cardiology	Adult population of the USA	male and female	1995	20–49	Diasorin 25-hydroxyvitamin D assay.	Using submaximal exercise test protocols.	0.1	Un adjusted
Książek	Poland	2016	Journal of Human Kinetics	43 Polish premier league soccer players	male and female	43	22.7	Electrochemil-uminescence (ECLIA) using the Elecsys system	During exercise testing with increasing loads were determined with a portable system K4 b2	0.02	Un adjusted
Pandey	USA	2018	The American Journal of Medicine	Older Patients with Heart Failure with Preserved Ejection Fraction	Male and female	112	70	Dia Sorin radioimmunoassay	Electronically braked cycle ergometer in the upright position	0.26	Season
Pandey	USA	2018	The American Journal of Medicine	37 healthy age-matched controls	Male and female	37	70	Dia Sorin radioimmunoassay	Electronically braked cycle ergometer in the upright position	0.077	Season

Table 1.
Characteristics of the included observational studies.

First author	Location	Population study	Subjects in/pl.*	Mean age in/pl	Mean bmi baseline in/pl	25(OH)D mean baseline in/pl	25(OH)D mean final in/pl	Duration in week	Dose vitamin D weekly (iu)	Mean change Vo2 max in/pl
Todd	UK	Healthy male and female athlete	22/20	20/20	23.89/22.31	44.49/40.93	81.77/46.33	12	21,000 Oral spray	_0.64/_2
Scholten	USA	Physically active males	14/14	32.8/29.9	23.4 /26.2	67.7/67.7	126.2/72.91	8	28,000Cap	_0.69/_0.29
Scholten	USA	Physically active males	6/6	32.2/30.3	23.1/25.9	87.92/75.66	118.24/79.64	8	28,000Cap	1.52/_0.86
Carrillo	USA	Overweight and obese adults during resistance training	10/13	26.2/26	30.6/31.9	20.8/18.1	33.4/23.5	12	28,000 CAP	6/5.7
Boxer	USA	Patients with heart failure (HF).	24/23	65.8/66	34.8/31.3	19.1/17.8		24	50,000 CAP	_0.17/_0.68
Singla	India	Adults with (T2DM) performed moderate-intensity aerobic exercise	9/9	41.5/39.3	28.1/27.9	9.8/10.9	38/10	12	60,000 CAP	_0.85/_8.63
Karefylakis	Sweden	Over weight/ obese men with vitamin D deficiency	17/18	49.8/49.4	31.5/31.2	44.3/44.2	70.5/49.8	24	14,000 Drop	0.4/0.6

In/pl.: intervention/placebo.

Table 2.
Characteristics of included randomized controlled trials.

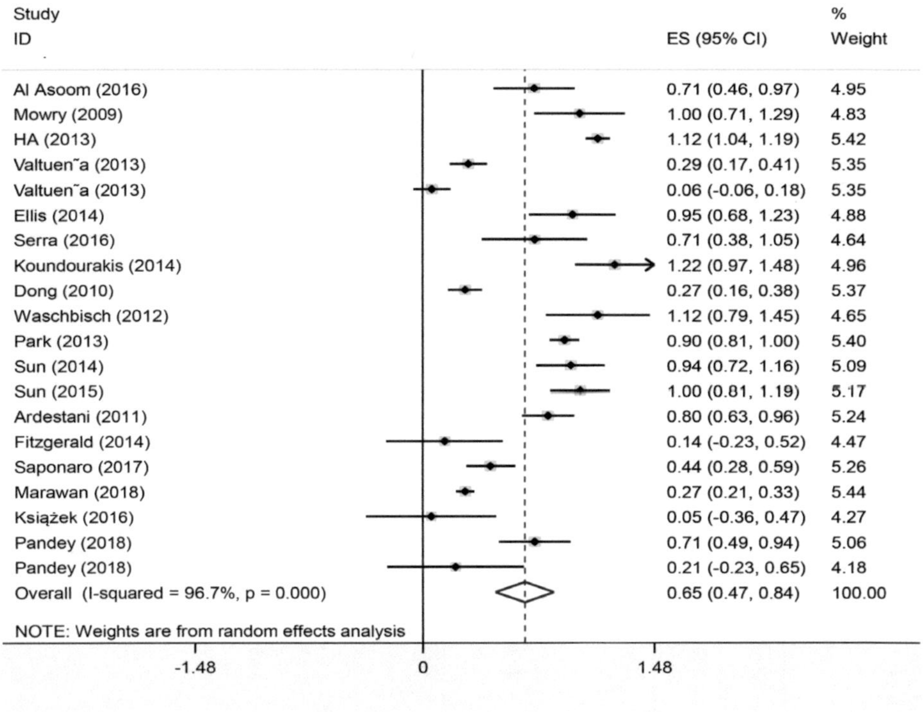

Figure 1.
Forest plot of correlation between 25(OH) D and CRF.

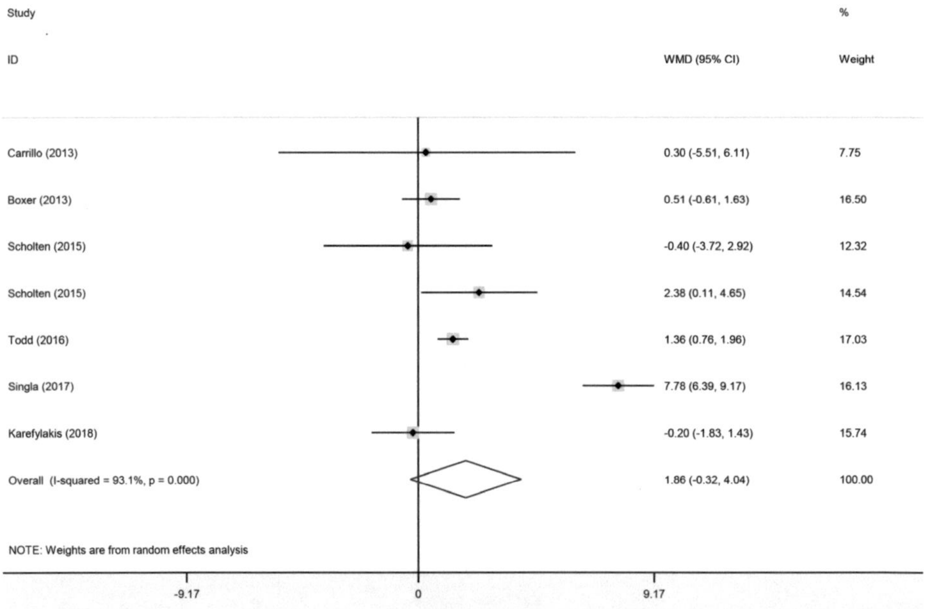

Figure 2.
Forest plot for the effects of vitamin D3 supplementation on CRF.

vitamin D deprivation can result in physiologic, histological, and electrophysiological alterations, supporting the role of vitamin D in maintaining muscle health. A strong resistance exercise's ability to restore strength may be predicted by a higher 25[OH] D concentration [29]. In younger adults, supplementing with vitamin D [4000 IU for 5 days] can increase muscle strength. However, it is important to note that the methodologies, dosages, participant characteristics, length of interventions, and findings of the research on the impact of vitamin D supplementation varied [30]. Both vitamin D insufficiency and vitamin D receptor [VDR] malfunction appear to have detrimental effects on the homeostasis of skeletal muscles. However, overexpression of VDR appears to have negative effects on skeletal muscle as well [31]. Although it has been demonstrated that vitamin D is also involved in the cellular metabolism of skeletal muscles, the precise molecular pathways that vitamin D activates in muscles are yet unknown. Vitamin D, through the activity of its active metabolite, 1,25[OH]2 D3, is crucial for normal calcium and phosphorus balance and the maintenance of skeletal health. The homeostasis of calcium involves vitamin D. Vitamin D controls the gut's absorption of calcium and maintains the levels of calcium and phosphate in the serum. It has been demonstrated to be crucial in controlling skeletal muscle tone and contraction [32]. The most recent study showed that treatment with 1,25[OH]2 D3 increased the oxygen consumption rate of skeletal muscle cells, demonstrating the role of vitamin D in the regulation of mitochondrial oxygen consumption and dynamics. An increase in respiration was associated with the production of ATP, suggesting that vitamin D improves the mitochondrial activity in muscle [33]. The hypothesis is that vitamin D can affect the blood supply to skeletal muscles and their ability to use oxygen due to the presence of the VDR in cardiac muscle, vascular tissue, and skeletal muscle. However, direct 1,25[OH]2 D3 administration of isolated mitochondria failed to increase oxygen consumption rate, indicating that 1,25[OH]2 D3's effects on oxygen consumption rate may be dependent on VDR or other extra-mitochondrial metabolic processes. People with lower vitamin D concentration get more benefits from vitamin D supplementation and more improvement in muscle strength [31, 33].

As a complementary basis to further strengthen the possible effect of vitamin D on muscle strength, numerous in vivo and in vitro experimental studies have demonstrated physiologic, histological, and electrophysiological alterations of skeletal muscle in severe vitamin D insufficiency, indicating a possible role for vitamin D in maintaining healthy muscles. As stated previously, it seems that the binding of vitamin D to its receptors promotes the absorption of inorganic phosphate needed for the production of energy-rich phosphate compounds [ATP] required for muscle cell contractility [34]. Additionally, high parathyroid hormone [PTH] has been proven to accelerate the breakdown of muscle proteins, and low vitamin D levels have been linked to secondary hyperparathyroidism [35]. Studies on muscle biopsies and electrophysiological tests further show the role of vitamin D in muscle cell activity. Treatment with vitamin D has been shown to reverse these changes, including an increased number of type II muscle fibers. Vitamin D deficiency has been linked to the atrophy of type II muscle fibers as well as nonspecific histological abnormalities like fatty infiltration, interstitial fibrosis, and sarcolemmal nuclear proliferation, all linking to lower muscle strength. Electrophysiological studies have connected low vitamin D levels to abnormal patterns, such as reduced motor unit potential length and amplitude, greater percentages of polyphasicity, and no concomitant denervation evidence [36].

3. Conclusion

By calculating VO2 max, serum 25[OH]D is directly correlated with CRF. The most recent findings suggest that vitamin D supplementation may result in higher CRF improvements in men and younger adults. However, since science is constantly evolving and changing, and many facts are misunderstood or unknown, these findings might be modified over time. Furthermore, multiple lines of research have indicated that vitamin D supplementation has a positive impact on aged people's muscle function. An effect, however, is not always there as more studies demonstrating the absence of an effect than studies demonstrating positive benefits have been published. The lack of clear explanations for the discrepant findings is due to the fact that studies showing positive benefits from those showing no effect of an increased vitamin D level do not appear to share many common traits.

Conflict of interest

The authors declare no conflict of interest.

Acronyms and abbreviations

CVD	cardiovascular disease
CRF	cardiorespiratory fitness
25[OH]D	25-hydroxyvitamin D
UV	ultraviolet
BMI	body mass index
VDR	vitamin D receptor
IOM	Institute of Medicine
IU	international unit

Author details

Amin Mirrafiei, Mahsa Firouzi, Nadia Babaee, Samira Davarzani and Sakineh Shab-Bidar*
Department of Community Nutrition, School of Nutritional Sciences and Dietetics, Tehran University of Medical Sciences [TUMS], Tehran, Iran

*Address all correspondence to: s_shabbidar@tums.ac.ir

References

[1] Greer FR. Defining vitamin D deficiency in children: Beyond 25-OH vitamin D serum concentrations. Pediatrics. 2009;**124**(5):1471

[2] Muscogiuri G et al. Can vitamin D deficiency cause diabetes and cardiovascular diseases? Present evidence and future perspectives. Nutrition, Metabolism and Cardiovascular Diseases. 2012;**22**(2):81-87

[3] Ardestani A et al. Relation of vitamin D level to maximal oxygen uptake in adults. The American Journal of Cardiology. 2011;**107**(8):1246-1249

[4] Owens DJ et al. "A systems-based investigation into vitamin D and skeletal muscle repair, regeneration, and hypertrophy." American Journal of Physiology-Endocrinology and Metabolism. 2015

[5] Blair SN et al. Influences of cardiorespiratory fitness and other precursors on cardiovascular disease and all-cause mortality in men and women. JAMA. 1996;**276**(3):205-210

[6] Lauretani F et al. Age-associated changes in skeletal muscles and their effect on mobility: an operational diagnosis of sarcopenia. Journal of Applied Physiology. 2003;**95**(5):1851-1860

[7] Zoeller RF Jr, Riechman SE, Dabayebeh IM, Goss FL, Robertson RJ, Jacobs PL. Relation between muscular strength and cardiorespiratory fitness in people with thoracic-level paraplegia. Archives of Physical Medicine and Rehabilitation. 2005;**86**(7):1441-1446

[8] Bischoff HA et al. Muscle strength in the elderly: Its relation to vitamin D metabolites. Archives of Physical Medicine and Rehabilitation. 1999;**80**(1):54-58

[9] Eslami O, Shidfar F, Akbari-Fakhrabadi M. Vitamin D and cardiorespiratory fitness in the general population: A systematic review. International Journal for Vitamin and Nutrition Research. 2017;**87**(5-6):330-341

[10] Nair R, Maseeh A. Vitamin D: The "sunshine" vitamin. Journal of Pharmacology and Pharmacotherapeutics. 2012;**3**(2):118-126

[11] Holick MF. Environmental factors that influence the cutaneous production of vitamin D. The American Journal of Clinical Nutrition. 1995;**61**(3):638S-645S

[12] Diehl JW, Chiu MW. Effects of ambient sunlight and photoprotection on vitamin D status. Dermatologic Therapy. 2010;**23**(1):48-60

[13] Youl PH, Janda M, Kimlin M. Vitamin D and sun protection: The impact of mixed public health messages in Australia. International Journal of Cancer. 2009;**124**(8):1963-1970

[14] Libon F, Cavalier E, Nikkels A. Skin color is relevant to vitamin D synthesis. Dermatology. 2013;**227**(3):250-254

[15] Wortsman J, Matsuoka LY, Chen TC, Lu Z, Holick MF. Decreased bioavailability of vitamin D in obesity. The American Journal of Clinical Nutrition. 2000;**72**(3):690-693

[16] Carlberg C. Nutrigenomics of vitamin D. Nutrients. 2019;**11**(3):676

[17] Jones K. Maitake: A potent medicinal food. Alternative and Complementary Therapies. 1998;**4**(6):420-429

[18] Robinson S, Canavan M, O'Donnell M, Mulkerrin E. Vitamin D supplementation-clarity required regarding treatment regimens and target plasma levels. QJM: An International Journal of Medicine. 2014;**107**(4):327-329

[19] Garg AX, Hackam D, Tonelli M. Systematic review and meta-analysis: When one study is just not enough. Clinical Journal of the American Society of Nephrology. 2008;**3**(1):253-260

[20] Hawkins MN, Raven PB, Snell PG, Stray-Gundersen J, Levine BD. Maximal oxygen uptake as a parametric measure of cardiorespiratory capacity. Medicine and Science in Sports and Exercise. 2007;**39**(1):103-107

[21] Beltz NM, Gibson AL, Janot JM, Kravitz L, Mermier CM, Dalleck LC. Graded exercise testing protocols for the determination of VO2max: Historical perspectives, progress, and future considerations. Journal of Sports Medicine (Hindawi Publ Corp). 2016;**2016**:3968393

[22] Min B. Effects of vitamin D on blood pressure and endothelial function. The Korean Journal of Physiology & Pharmacology. 2013;**17**(5):385

[23] D'Silva A et al. Cardiovascular remodeling experienced by real-world, unsupervised, young novice marathon runners. Frontiers in Physiology. 2020;**11**:232

[24] Moreira-Pfrimer LD, Pedrosa MA, Teixeira L, Lazaretti-Castro M. Treatment of vitamin D deficiency increases lower limb muscle strength in institutionalized older people independently of regular physical activity: A randomized double-blind controlled trial. Annals of Nutrition and Metabolism. 2009;**54**(4):291-300

[25] Girgis CM, Clifton-Bligh RJ, Hamrick MW, Holick MF, Gunton JE. The roles of vitamin D in skeletal muscle: Form, function, and metabolism. Endocrine Reviews. 2013;**34**(1):33-83

[26] Marawan A, Kurbanova N, Qayyum R. Association between serum vitamin D levels and cardiorespiratory fitness in the adult population of the USA. European Journal of Preventive Cardiology. 2019;**26**(7):750-755

[27] Karefylakis C, Särnblad S, Ariander A, Ehlersson G, Rask E, Rask P. Effect of vitamin D supplementation on body composition and cardiorespiratory fitness in overweight men—a randomized controlled trial. Endocrine. 2018;**61**(3):388-397

[28] Distefano G, Goodpaster BH. Effects of exercise and aging on skeletal muscle. Cold Spring Harbor Perspectives in Medicine. 2018;**8**(3)

[29] Polly P, Tan TC. The role of vitamin D in skeletal and cardiac muscle function. Frontiers in Physiology. 2014;**5**:145

[30] Hossein-nezhad A, Holick MF. Vitamin D for health: A global perspective. Mayo Clinic Proceedings (Elsevier). Jul 2013;**88**(7):720-755. DOI: 10.1016/j.mayocp.2013.05.011

[31] Dzik KP, Kaczor JJ. Mechanisms of vitamin D on skeletal muscle function: Oxidative stress, energy metabolism and anabolic state. European Journal of Applied Physiology. 2019;**119**(4):825-839

[32] Gil Á, Plaza-Diaz J, Mesa MD. Vitamin D: classic and novel actions. Annals of Nutrition and Metabolism. 2018;**72**(2):87-95

[33] Latham CM et al. Vitamin D promotes skeletal muscle regeneration and mitochondrial health. Frontiers in Physiology. 2021;**12**:660498

[34] Mosekilde L. Vitamin D and the elderly. Clinical Endocrinology. 2005;**62**(3):265-281

[35] Ritz E, Boland R, Kreusser W. Effects of vitamin D and parathyroid hormone on muscle: Potential role in uremic myopathy. The American Journal of Clinical Nutrition. 1980;**33**(7):1522-1529

[36] Rejnmark L. Effects of vitamin D on muscle function and performance: A review of evidence from randomized controlled trials. Therapeutic Advances in Chronic Disease. 2011;**2**(1):25-37

Chapter 11

Pharmacological Efficacy and Mechanism of Vitamin D in the Treatment of "Kidney-Brain" Disorders

Jia-Li Zhang, Yong-Jun Wang and Yan Zhang

Abstract

Accumulating evidences have shown that serum 25-hydroxyvitamin D concentrations were inversely correlated with the incidence or severity of coronavirus disease 2019 (COVID-19), caused by severe acute respiratory syndrome coronavirus 2 (SARS-CoV-2), and that vitamin D deficiency might be associated with an increased susceptibility to many of the complications accompanied by COVID-19, such as disorders in kidney and brain. Our previous experimental studies demonstrated that vitamin D and its analogs could protect from kidney diseases, neuroinflammation, and musculoskeletal disorders such as osteoporosis and muscle atrophy, through the suppressive effects on overactivation of the renin-angiotensin system (RAS) in tissues. Moreover, we published a review describing the therapeutic effects of traditional Chinese medicine (TCM) for organ injuries associated with COVID-19 by interfering with RAS. In the TCM principle "Kidney dredges brain," this chapter will emphasize the potential preventive and therapeutic effects of vitamin D on both renal injuries and central nervous system disorders in COVID-19 patients and further elucidate the pharmacological effects with underlying mechanisms of vitamin D in "Kidney-Brain" disorders.

Keywords: vitamin D receptor, kidney, brain, renin-angiotensin system, traditional Chinese medicine, vitamin D

1. Introduction

The outbreak of Coronavirus Disease 2019 (COVID-19) has created a global public health crisis. Observational studies provided evidence that serum 25-hydroxyvitamin D [25(OH)D] concentration was inversely correlated with the incidence or severity of COVID-19 [1]. Moreover, very severe vitamin D deficiency (<10 ng/ml) was considerably more common in COVID-19 patients than in non-COVID-19 ones [2]. Consistently, a significant correlation between vitamin D sufficiency and reduction in clinical severity and inpatient mortality from COVID-19 disease has been explored [3, 4].

Actually, as vitamin D is concerned, traditional Chinese medicine (TCM) and Western medicine could share similar philosophical logic to fight against COVID-19, mainly because in TCM theory, the pathogenesis of COVID-19 is closely associated with cold dampness, which could be attenuated by sun exposure and Wen-Yang herbs, both of which could restore the blood level of vitamin D in Western medicine [5]. Clinically, TCM medications have been exhibiting benefits in decreasing the rate of disease progression, time to the resolution of fever, and rate of progression to severe COVID-19 cases [6], and we published a review summarizing the pharmacological interventions and the underlying mechanisms of TCM for organ injuries associated with COVID-19 [7].

As we know, the renal 1α-OHase enzyme catalyzes the biosynthesis of active vitamin D, $1,25(OH)_2D_3$, and conversely, the 24-OHase enzyme in the kidney deactivates vitamin D via hydroxylation at site 24 on the chemical structure of 25(OH)D and $1,25(OH)_2D_3$. In our group, we have published a series of research articles uncovering that the kidney-tonifying herb Fructus Ligustri Lucidi could manage vitamin D metabolism and enhance circulating $1,25(OH)_2D_3$ level [8–10]. Intriguingly, there are TCM theories supporting the relevance between kidney and brain, such as "Interaction between Kidney and Brain," and "Kidney dominates bone, and dredges brain," *etc.*, where the biological basis for this interaction might be attributed to vitamin D. Therefore, the effects and the mechanisms of vitamin D on comorbidity in kidney and brain of COVID-19 patients are elaborated in this chapter.

2. Vitamin D and kidney injuries

2.1 Clinical practice

Several previous studies have demonstrated that the risk of COVID-19 and associated death increases with the coexistence of various underlying diseases, including liver and kidney failure, cerebrovascular disease, chronic obstructive pulmonary disease, coronary heart disease, hypertension, diabetes, and so forth [11–16]. Among those comorbid diseases, the incidence of kidney injuries in the general population after infection with SARS-CoV-2 was around 3–15%, 14.5–50% in patients with severe COVID-19 infection in the intensive care unit, and even higher in patients with chronic kidney disease (CKD), which is related to severe infection and higher fatality rate in COVID-19 patients [17, 18]. In a retrospective case-control study from a Los Angeles Health System, Chang *et al.* [19] observed that renal diseases were the predominant premorbid risk factors for COVID-19 patients. In New York, a longitudinal prediction study at two centers showed that the death rate in COVID-19 patients accompanied by acute kidney injury (AKI) is approximately five times that of mortality in non-AKI patients (31.5% and 6.9%) and that the COVID-19 patients developed AKI about 3 days after hospitalization [20].

A growing body of evidence suggests that vitamin D and COVID-19 are linked. The first study to examine whether the last vitamin D status before COVID-19 testing is associated with COVID-19 test results of 489 patients was published after the outbreak of the COVID-19 pandemic. This single-center retrospective cohort study concluded that adults with hypovitaminosis D were more likely to be infected by SARS-CoV-2 [21]. Similarly, in 20 European countries, substantial inverse associations between mean blood 25(OH)D concentrations and the frequency of COVID-19 cases and mortality were discovered [22]. The above data illustrate the close correlation between serum vitamin D levels and the risk rate of developing COVID-19. In addition, at the same time, studies found that vitamin D supplementation could reduce the risk of being infected with SARS-CoV-2. As presented in a retrospective cohort study done in Switzerland, vitamin D supplementation reduced the probability of SARS-CoV-2 infections [23]. Furthermore, frequent vitamin D_3 supplementation, at least in the elderly, in boluses taken routinely throughout the year preceding diagnosis, has indicated a reduction in the risk of mortality and clinical improvement in old COVID-19 patients [24]. In a short term, randomized, placebo-controlled trial in 25(OH)D deficient (<20 ng/mL) COVID-19 individuals from India, 62.5% of those treated with 60,000 IU/1500 μg/day of vitamin D_3 for 7–14 days were negative for SARS-CoV-2 after 21 days, compared with just 20.8% of those who were not given vitamin D_3 [25].

Vitamin D is a vital protector for inhibiting inflammation and cytokine storms in the kidney [26]. The correlations were demonstrated between low vitamin D levels and the risk of influenza infection. Same as influenza, different studies showed that vitamin D status could influence the outcome of COVID-19 patients, including kidney injuries [27]. In Spain, a retrospective cohort clinical trial was held to compare whether the administration or not of oral calcifediol could alleviate mortality risk and the underlying diseases arising from COVID-19 [24]. Among the 537 included COVID-19 patients, those who received calcifediol (0.266 mg/capsule, two capsules on entry, and then one capsule on days 3, 7, 14, 21, and 28) were more likely to have a low rate of CKD and even mortality [24]. The COVID-19 patients accompanied by CKD with maintenance hemodialysis have a very high 3-month mortality rate, but researchers found that the same type of patient treated with active vitamin D had a lower risk of mortality caused by COVID-19 [28]. The facts all indicated that either serum vitamin D status or vitamin D supplementation has a strong link with the degree of severity of kidney injuries associated with COVID-19.

2.2 Mechanism studies

2.2.1 Renin-angiotensin system (RAS)

SARS-CoV-2 enters cells when its spike proteins are bound to angiotensin-converting enzyme 2 (ACE2) receptors, which are the potent negative regulators on the RAS and are highly expressed in the kidney [7]. The excess activity of the renal RAS, characterized as the increased production of angiotensin II (Ang II), is responsible for kidney destruction, inflammation, and functional failure related to SARS-CoV-2 [29].

Vitamin D inhibits renin expression and in turn reduces Ang II expression, thus, serving as a negative RAS regulator [29, 30]. The deficiency of vitamin D activates the intrarenal RAS, thereby inducing an increase in the level of Ang II, which is an important stimulator of kidney injury [31, 32]. Our study demonstrated that active vitamin D analogs paricalcitol and doxercalciferol were able to suppress RAS activation, alleviate glomerular and tubulointerstitial damage, and reduce proteinuria in streptozotocin (STZ, 40 mg/kg)-induced diabetic DBA/2 J mice [33, 34]. Similarly, treatment of STZ (60 mg/kg)-induced type 1 diabetic rats with calcitriol (0.2 μg/kg, i.g.) significantly reduced urine albumin and improved glomerular ultrastructure by reducing the renin expression and alleviating the oxidative stress of the kidneys [35]. The role of RAS in the kidney of type 2 diabetic mice (db/db mice) in our study was consistent with those studies performed on the type 1 diabetic animal models [36, 37]. As vitamin D exerts a vital effect by binding to vitamin D receptor (VDR), which is widely expressed in various organs and tissues including kidneys, we considered that VDR signaling may be a paramount modulator in the process of kidney injuries and therefore constructed the VDR knockout mice and performed a series of systematic studies. At first, our study found a significant elevation in renin gene expression in VDR-null mice [38, 39]. In *in vitro* experiments, we revealed that the direct inhibitory effect of $1,25(OH)_2D_3$ on renin was mainly attributed to its ability to inhibit the transcription of the renin gene and consequently cut off the activation of the RAS [38, 39]. The results of our subsequent study revealed that the absence of VDR stimulated severe renal damage with marked tubular atrophy, interstitial fibrosis, and increased renin expression and Ang II accumulation in the obstructed kidney of mice after surgery of unilateral ureteral obstruction (UUO) [29].

In line with the *in vivo* studies, in the podocytes with the absence of VDR, the mRNA levels of angiotensinogen (AGT), renin, and Ang II type 1 receptor (AT1R) were significantly upregulated, displaying the activation of RAS and therefore exacerbating podocytes damage [40]. Vitamin D and its analogs also repressed the activation of RAS in other renal cells, such as HIV-induced tubular cell injury, high-glucose-induced mesangial cells, juxtaglomerular cells, and Ang II-induced primary tubular cells [41, 42].

Collectively, vitamin D might prevent kidney injury associated with SARS-CoV-2 infection by attenuating renal RAS as shown by an upregulation of ACE2 expression and downregulation of renin expression as well as a reduction in the production of Ang II locally in the kidney.

2.2.2 Epithelial-mesenchymal transition (EMT)

Researchers supported that in addition to the RAS imbalance caused by SARS-CoV-2 infection, the COVID-19 may also bring about the EMT, which has been reported as a major mechanism responsible for the abnormal accumulation of extracellular matrix (ECM). As reported, the accumulation of proteins and fibroblasts in ECM is a predominant factor in causing most kidney diseases [29]. It is believed that vitamin D could prevent kidney fibrosis by repressing the process of EMT [29].

As shown in recent research, calcitriol and paricalcitol (at equivalent doses of 1000 IU/kg) prevented the renal fibrosis in the 7/8 nephrectomy model within 4 weeks of treatment through the inhibition of EMT characterized by the changes of E-cadherin and Snail [43]. Similarly, type I and type III collagen, fibronectin, α-smooth muscle actin, and E-cadherin, which are the typical markers of EMT, were significantly regulated in UUO mice treated with paricalcitol, which therefore ameliorated renal interstitial fibrosis and preserved tubular epithelial integrity in obstructive nephropathy [44]. Furthermore, the *in vitro* experimental studies demonstrated that the isolated primary tubular cells from VDR-null mice showed a significant EMT process, which was reached by analyzing the abundance of landmark indicators of EMT [29]. Consistent with the conclusion drawn from the primary tubular cell study, paricalcitol treatment also profoundly suppressed EMT in the human renal proximal tubular epithelial cells [45].

Overall, EMT might be one of the key pathogenic pathways for COVID-19-induced kidney injury, and the inhibition of EMT by vitamin D analogs suggests that it may ameliorate renal injury *via* declining renal EMT caused by SARS-CoV-2 infection.

2.2.3 Oxidative stress, inflammation, and cytokine storm

As mentioned earlier, vitamin D is not only an essential factor for modulating real RAS and suppressing the EMT process, but also for regulating oxidative stress and inhibiting inflammation and cytokine storm, consequently reducing COVID-19-induced kidney damage [26]. The SARS-CoV-2 infection triggers the massive production of reactive oxygen species (ROS) and promotes oxidative damage. Jain *et al.* [46] have proposed that ROS overproduction and excessive oxidative stress are responsible for impaired immunity, cytokine storm secretion, and the onset of organ dysfunction in response to COVID-19 infection. The insights could be interpreted as follows: for example, Wu *et al.* [47] revealed that cholecalciferol has the potential, as a clinical drug, to protect renal function in ischemia/reperfusion (I/R)-induced AKI by reducing ROS production and inhibiting oxidative stress. Other studies also suggested that vitamin D analogs could participate in the induction of intracellular free radical scavenging and attenuate kidney disorder through the downregulation of mTOR expression and autophagy-related oxidant response [48–50].

Low 25(OH)D status in COVID-19 patients was correlated with high levels of interleukin-6 (IL-6) and C-reactive protein (CRP), which are the independently inflammatory markers. Furthermore, the COVID-19 patients with insufficient 25(OH)D content may exert a high incidence of inflammation-induced renal injury [51]. Several experimental studies have reported that the administration of VDR activators reduced the presence of inflammatory cells in the kidney, thereby suppressing inflammatory responses and cytokine storms [7, 52–54]. Additionally, vitamin D intervention could decrease the production of inflammatory cytokines such as IL-6, IL-8, IL-12, IL-17, tumor necrosis factor-α (TNF-α), and interferons-γ (IFN-γ), and thus prevent inflammation from progressing and damaging other organs, including the kidneys [55–57]. As a result, numerous preclinical studies have been conducted using vitamin D as a treatment for various types of AKI, such as sepsis-induced AKI, with promising results in mitigating both renal oxidative stress and the expression of inflammatory cytokines in kidney [58].

Therefore, vitamin D could have the potential in diminishing the cytokine storm caused by COVID-19 and could exert protective effects against kidney injury.

2.2.4 Immune response

The active vitamin D molecule 1,25$(OH)_2D_3$ could be produced in the kidneys and in extrarenal tissues such as activated monocytes/macrophages, where VDR is also expressed and is therefore vitamin D targets as well [59, 60]. Various studies have shown a stimulatory effect of vitamin D on Tregs ($CD4^+$, $CD25^+$, $CD127^-$, $FoxP3^+$), which are the important immune response cells in humans [61–63].

In detail, a study by Yuan *et al.* [64] found that combined treatment with vitamin D and tacrolimus effectively alleviated renal tissue damage in rats with IgA nephropathy through modulating the immune response and the nuclear factor kappa-B/toll-like receptor 4 (NF-κB/TLR4) pathway, the overactivation of which is a typical appearance of immune injury. Similar findings have been reported that 1,25$(OH)_2D_3$ protected against tubulointerstitial fibrosis by downregulating the innate immune NF-κB/TLR4 pathway in STZ (45 mg/kg)-induced diabetic rats with kidney disease and in high-glucose (25 mmol/L)-induced NRK-52E cells (a rat kidney tubular epithelial cell line) [65]. In addition, Penna *et al.* [66] demonstrated that the treatment of nonobese diabetes mice, a model of susceptible autoimmune disease, with a synthetic analog of 1,25$(OH)_2D_3$ could reduce IL-17 expression. Several studies also supported that vitamin D could play a role in the cross talk between innate and adaptive immunity in CKD patients, illustrating that vitamin D could improve organ damage, including kidney damage, by modulating the immune response [67–69].

Since SARS-CoV-2 infection affects the immune system first and foremost, vitamin D intervention could somehow regulate the body's immune function and the stress of immune cells in the kidney. Hence, it could be assumed that the modulations on the immune response might be one potential mechanism for the beneficial effects of vitamin D on kidney deterioration.

3. Vitamin D and CNS disorders

3.1 Clinical practice

As the current understanding of COVID-19 continues to evolve, accumulating evidence demonstrated the neurological impact of this novel virus [70], particularly, the term "NeuroCovid" has been proposed in 2020 [71]. During the acute phase of COVID-19, about 36% of cases developed neurological symptoms of which 25% could be attributed to the direct involvement of the central nervous system (CNS) [72]. Patients with neurological deficits such as Parkinson's disease (PD) did not exhibit an elevation in COVID-19 risk and mortality compared with the general population [73, 74]; however, COVID-19 might lead to the medium- and long-term consequences on CNS with neurodegenerative and neuropsychiatric diseases manifested as depression, insomnia, cognitive decline, accelerated aging, Parkinson's disease (PD), or Alzheimer's disease [71, 75]. The infection with SARS-CoV-2 even aggravates the CNS disorders and neurological complications of COVID-19 patients with preexisting neurological injury. In children with multiple sclerosis, the results of the web-based survey showed high anxiety levels during the pandemic [76]. Additionally, the affected patients associated with cognitive deficits might be at higher risk of cognitive decline after overcoming the COVID-19 infection [70]. Importantly, a systematic review of studies reporting data on PD patients with a diagnosis of COVID-19 indicated a higher case fatality in PD patients affected by COVID-19 than the general population [74]. Therefore, a strengthened awareness of the possibility of neurological involvement and a further investigation into the relevant pathophysiology would be essential to understand and ultimately abrogate SARS-CoV-2-related neurological symptoms [77].

An unselected large cohort study from Italy showed that the nonadvanced PD patients without vitamin D supplementation were more likely to be infected [73], and a retrospective survey from Spain elucidated that lower blood level of vitamin D was one of the main factors for developing COVID-19 in children with neuroimmunological disorders [78]. Consistently the systematic analysis including 16 studies reporting on a total of 11,325 PD patients suggested vitamin D might be a key protective factor against this infection [74], and the meta-analysis indicated the close correlation of vitamin D supplementation with COVID-19 in people with PD [79]. Furthermore, an early study using a multivariate general linear model found that a low serum level of 25(OH)D could predict an increased vulnerability to the stressful impact of the COVID-19 outbreak [80]. Collectively, vitamin D deficiency in circulation not only increases susceptibility to COVID-19 in patients with CNS disorders but also accelerates or aggravates preexisting neurodegenerative disease in COVID-19 patients.

Most of the emerging clinical results supported the beneficial effects of vitamin D supplements or therapy on neurological complications in COVID-19 patients, in accordance with the neuroprotective effects of vitamin D and its analogs. It is well elucidated that SARS-CoV-2 is a neuroinvasive virus capable of eliciting a cytokine storm, with persistent effects in specific populations. The impact of SARS-CoV-2 infection on the onset and progression of neurological diseases of neuroinflammatory origin is regarded as the potential cause of a delayed pandemic [81]. Remarkably, as a nonclassical role beyond action on skeletal homeostasis, the pharmacological regulations of vitamin D on inflammation responses including neuroinflammation have been widely studied. An interesting review stated that vitamin D could partially produce positive effects on the development of brain function for infants of mothers who experienced viral infections in early pregnancy by reducing some pro-inflammatory cytokines [82]. Vitamin D might act as a strong immunosuppressant repressing cytokine release syndrome in COVID-19 via attenuating the production and secretion of crucial pro-inflammatory cytokines including NF-kB, IL-6, IL-1β, and TNF [83]. One recent review implicated that the immunomodulatory effects of vitamin D significantly reduced the level of pro-inflammatory interleukins and enhanced the synthesis of anti-inflammatory chemical mediators [84]. Taken together, supplementation with vitamin D could be an effective option to avoid the development and progression of neurodegenerative pathologies in post-COVID-19 patients.

3.2 Mechanism studies

Given the extrarenal regulation of vitamin D on tissue function, its extrarenal metabolism, especially in CNS, will be extremely concerned in the research studies on neurological illnesses accompanied by COVID-19. Experimental data showed that VDR is expressed in CNS such as neurons and microglia, and $25(OH)D_3$ could be directly metabolized to $1,25(OH)_2D_3$ due to the local presence of 1α-hydroxylase, implying a potential modulation of vitamin D in CNS in an autocrine or paracrine fashion. $1,25(OH)_2D_3$ could stimulate the expression of glial cell line–derived neurotrophic factor, nerve growth factor, and neurotrophins-like nerve growth factor (NGF), thereby preventing loss of neural sensation in COVID-19 [83]. Moreover, $1,25(OH)_2D_3$ could promote the expression of brain-derived neurotrophic factor (BDNF), neurotrophin-3 (NT3), and neurotrophin receptor $p75^{NTR}$ in neurons, glial cells, and Schwann cells [83], as well as induce the migration and differentiation of oligodendrocyte progenitors and enhance remyelination of neurons to improve neurotransmission in a model of toxic demyelination [85]. Additionally, it improved serotoninergic and dopaminergic neurotransmission in cultured neuronal cells by modulating serotonin and dopamine metabolism [86]. These effects account for the potential therapeutic efficacy of vitamin D on COVID-19-derived neuropsychiatric disorders [84].

Considering our previous work emphasized the role of RAS in the development of tissue injuries and the inhibitory effects of vitamin D on overactivity of tissue RAS, we attempted to uncover the underlying molecular mechanisms involved in the protection of vitamin D in COVID-19 patients from CNS damages on the aspect of brain RAS, which has been proposed five decades ago [87]. Human studies on the postmortem brain revealed that human coronavirus variants and SARS-CoV-2 could infect neurons and glia, demonstrating that SARS-CoV-2 may have similar neurovirulence [77]. In fact, the SARS-CoV-2 virus could use the ACE2 to cross the blood-brain barrier and invade neuronal and glial cells, as the studies have explored that SARS-CoV-2 has a high affinity for its receptor, the ACE2 protein [84, 88]. Furthermore, the research data showed the expression of ACE2 in neuronal and glial cells [89], which are also potentially vulnerable to SARS-CoV-2 infection. Attractively, a few studies have demonstrated the existence of RAS components in the basal ganglia, and particularly in the nigrostriatal system [90], even in mitochondria of dopaminergic neurons [91], though there are still controversial opinions about the presence of brain RAS as the angiotensin generation in the brain is concerned [87, 92].

It is well defined that there are two counterregulatory arms within RAS, namely the classical axis ACE/Ang II/AT1R and the newly emerged axis ACE2/Ang(1–7)/Mas [93]. The identification of the ACE homolog, ACE2 as a key Ang(1–7)-forming enzyme, unravels the existence of a distinct enzymatic pathway for the production of Ang(1–7), which has a broad range of effects in different organs and tissues that goes beyond its initially described cardiovascular and renal actions [94]. The decline in ACE2 expression that occurs with aging has been associated with higher morbidity and mortality rates in older adults [95]. Furthermore, numerous studies discovered that the cross talk and the interaction between the dual-axis systems of RAS contribute to tissue homeostasis. Our research project entitled "Biological effect of the double axes within RAS, ACE/Ang II/AT1R and ACE2/Ang(1-7)/Mas, in bone metabolism disturbance induced by high glucose and intervention study of active components in kidney-tonifying TCM," funded by National Natural Science Foundation of China, illustrated that the two axes distinctly regulated the differentiation and functions of osteoblasts and osteoclasts upon exposure to high glucose [96]. Our study [96] and another study [97] support the concept that the ACE2/Ang(1–7)/Mas axis is able to counteract most of the deleterious actions of the ACE/Ang II/AT1R axis, especially in pathological conditions. Thus, we suppose that the interfering of SARS-CoV-2 with ACE2 in the brain would lead to a disturbance between the two axes and, in turn, produce deleterious effects in CNS observed in infected patients.

In vivo and in vitro studies clarified a counterregulatory interaction between dopamine and angiotensin receptors [98] and between SIRT3 and angiotensin receptors [99] in the striatum and substantia nigra, especially in an age-dependent manner, thereafter induced dopaminergic function injury accounting for the rise in the risk of neurodegenerative diseases, such as PD. Besides that, the hyperactivation of the ACE/Ang II/AT1R axis could exacerbate dopaminergic cell death, the animal study explicated that the Ang(1–7)/Mas axis possessed a neuroprotective role in the dopaminergic system, and in turn, ameliorated aging-related vulnerability to neurodegeneration [100].

Vitamin D could raise the bioavailability and upregulate the expression of ACE2, which may be responsible for trapping and inactivating SARS-CoV-2 [101, 102]. Importantly, vitamin D could mitigate the RAS-activation-evoked tissue destruction by serving as an RAS inhibitor. The overall effects of vitamin D on brain RAS are assumed as a drop-in Ang II level and a rise in Ang(1–7) level by inducing the ACE2/Ang(1–7)/Mas axis activity and suppressing ACE/Ang II/AT1R axis [88, 103]. Our research articles have reported that active vitamin D analog paricalcitol could dramatically improve LPS-induced depressive-like behavior of mice by abolishing neuroinflammation via diminishing RAS activity in the hypothalamus [104], and the kidney-tonifying traditional herb Fructus Ligustri Lucidi displayed the suppressive effects on levels of pro-inflammatory cytokines by improving vitamin D metabolism [105]. Consistent with these findings, vitamin D supplementation appeared to reverse COVID-19-related neurodegeneration and neuroinflammation, which are aggravated in Parkinson's and Alzheimer's patients [95]. These pieces of evidence heighten the key role of vitamin D as a neuroprotective and neuroreparative agent against the neurological sequelae of COVID-19.

Taken together, the mechanism studies revealed the crucial role of VDR in the protection of nephropathy through regulating multiple biological events (**Figure 1**) and that

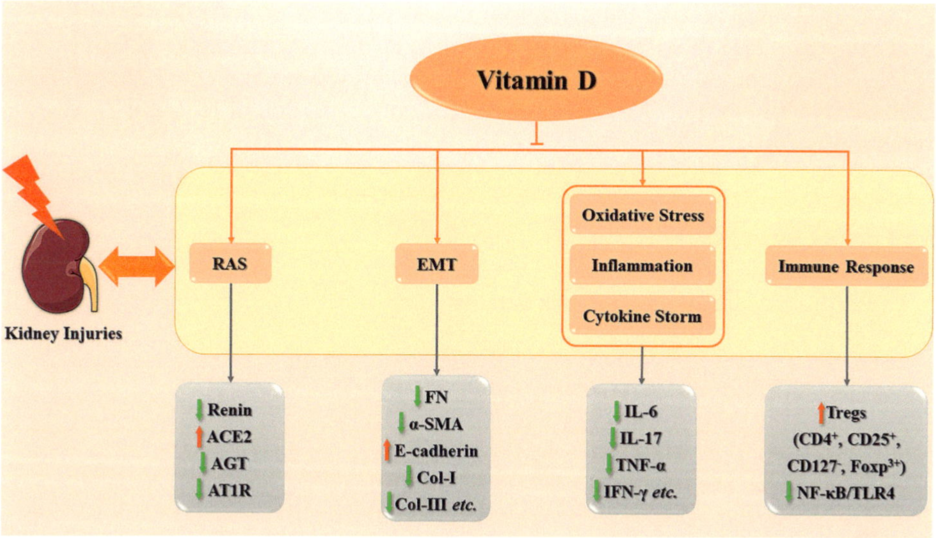

Figure 1.
Vitamin D displayed nephroprotective effects through regulating multiple biological events.

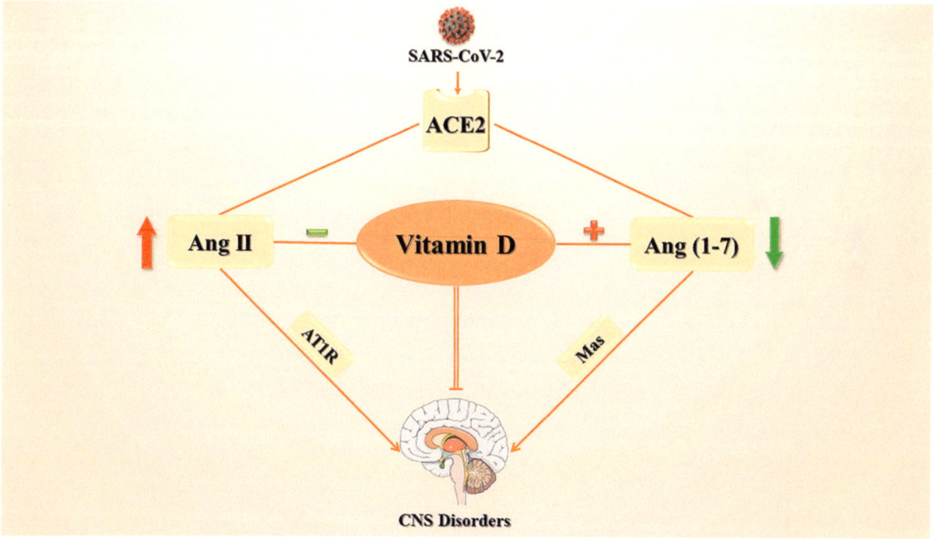

Figure 2.
Vitamin D exerted neuroprotective effects via, at least partially, modulating RAS homeostasis in CNS.

vitamin D exerted neuroprotective effects by balancing RAS in CNS (**Figure 2**), thereby vitamin D and its analogs possess the high potential in the protection and treatment of kidney and CNS disorders associated with COVID-19.

Acknowledgements

This chapter was supported in part by National Natural Science Foundation of China (82074468), Scientific and Innovative Action Plan from Science and

Technology Commission of Shanghai Municipality (21400760400), National Key R&D Program (2018YFC1704302) and Program for Innovative Research Team (2015RA4002) from Ministry of Science and Technology of China, and Shanghai Collaborative Innovation Center of Industrial Transformation of Hospital TCM Preparation.

Conflict of interest

The authors declare no conflict of interest.

Author details

Jia-Li Zhang, Yong-Jun Wang and Yan Zhang*
Shanghai University of Traditional Chinese Medicine, Longhua Hospital, Shanghai, China

*Address all correspondence to: medicineyan@aliyun.com

References

[1] Mercola J, Grant WB, Wagner CL. Evidence regarding vitamin D and risk of COVID-19 and its severity. Nutrients. 2020;**12**(11):3361. DOI: 10.3390/nu12113361

[2] Demir M, Demir F, Aygun H. Vitamin D deficiency is associated with COVID-19 positivity and severity of the disease. Journal of Medical Virology. 2021;**93**(5):2992-2999. DOI: 10.1002/jmv.26832

[3] Papadimitriou DT, Vassaras AK, Holick MF. Association between population vitamin D status and SARS-CoV-2 related serious-critical illness and deaths: An ecological integrative approach. World Journal of Virology. 2021;**10**(3):111-129. DOI: 10.5501/wjv.v10.i3.111

[4] Ali N. Role of vitamin D in preventing of COVID-19 infection, progression and severity. Journal of Infection and Public Health. 2020;**13**(10):1373-1380. DOI: 10.1016/j.jiph.2020.06.021

[5] Zhao F, Yang Z, Wang N, et al. Traditional Chinese medicine and western medicine share similar philosophical approaches to fight COVID-19. Aging & Disease. 2021;**12**(5):1162-1168. DOI: 10.14336/AD.2021.0512

[6] Jeon SR, Kang JW, Ang L, et al. Complementary and alternative medicine (CAM) interventions for COVID-19: An overview of systematic reviews. Integrative Medicine Research. 2022;**11**(3):100842. DOI: 10.1016/j.imr.2022.100842

[7] Zhang JL, Li WX, Li Y, et al. Therapeutic options of TCM for organ injuries associated with COVID-19 and the underlying mechanism. Phytomedicine. 2021;**85**:153297. DOI: 10.1016/j.phymed.2020.153297

[8] Zhang Y, Dong XL, Leung PC, et al. Fructus Ligustri Lucidi extract improves calcium balance and modulates the calciotropic hormones level and vitamin D-dependent gene expression in aged ovariectomized rats. Menopause. 2008;**15**(3):558-565. DOI: 10.1097/gme.0b013e31814fad27

[9] Zhang Y, Lai WP, Leung PC, et al. Improvement of Ca balance by Fructus Ligustri Lucidi extract in aged female rats. Osteoporosis International. 2008;**19**(2):235-242. DOI: 10.1007/s00198-007-0442-9

[10] Zhang Y, Diao TY, Wang L, et al. Protective effects of water fraction of Fructus Ligustri Lucidi extract against hypercalciuria and trabecular bone deterioration in experimentally type 1 diabetic mice. Journal of Ethnopharmacology. 2014;**158**(Part A):239-245. DOI: 10.1016/j.jep.2014.10.025

[11] Liu M, Mei K, Tan Z, et al. Liver fibrosis scores and hospitalization, mechanical ventilation, severity, and death in patients with COVID-19: A systematic review and dose-response meta-analysis. Canadian Journal of Gastroenterology and Hepatology. 2022;**2022**:7235860. DOI: 10.1155/2022/7235860

[12] Singh J, Malik P, Patel N, et al. Kidney disease and COVID-19 disease severity-systematic review and meta-analysis. Clinical and Experimental Medicine. 2022;**22**(1):125-135. DOI: 10.1007/s10238-021-00715-x

[13] Yu JN, Wu BB, Yang J, et al. Cardio-cerebrovascular disease is

associated with severity and mortality of COVID-19: A systematic review and meta-analysis. Biological Research for Nursing. 2021;**23**(2):258-269. DOI: 10.1177/1099800420951984

[14] Iaccarino G, Grassi G, Borghi C, et al. Age and multimorbidity predict death among COVID-19 patients: Results of the SARS-RAS study of the Italian society of hypertension. Hypertension. 2020;**76**(2):366-372. DOI: 10.1161/HYPERTENSIONAHA.120.15324

[15] Yang J, Zheng Y, Gou X, et al. Prevalence of comorbidities and its effects in patients infected with SARS-CoV-2: A systematic review and meta-analysis. International Journal of Infectious Diseases. 2020;**94**:91-95. DOI: 10.1016/j.ijid.2020.03.017

[16] Bradley SA, Banach M, Alvarado N, et al. Prevalence and impact of diabetes in hospitalized COVID-19 patients: A systematic review and meta-analysis. Journal of Diabetes. 2022;**14**(2):144-157. DOI: 10.1111/1753-0407.13243

[17] Naicker S, Yang CW, Hwang SJ, et al. The novel coronavirus 2019 epidemic and kidneys. Kidney International. 2020;**97**(5):824-828. DOI: 10.1016/j.kint.2020.03.001

[18] Hirsch JS, Ng JH, Ross DW, et al. Acute kidney injury in patients hospitalized with COVID-19. Kidney International. 2020;**98**(1):209-218. DOI: 10.1016/j.kint.2020.05.006

[19] Chang TS, Ding Y, Freund MK, et al. Prior diagnoses and medications as risk factors for COVID-19 in a Los Angeles health system. medRxiv. 9 Jul 2020. DOI: 10.1101/2020.07.03.20145581

[20] Lu JY, Hou W, Duong TQ. Longitudinal prediction of hospital-acquired acute kidney injury in COVID-19: A two-center study. Infection. 2022;**50**(1):109-119. DOI: 10.1007/s15010-021-01646-1

[21] Meltzer DO, Best TJ, Zhang H, et al. Association of vitamin D status and other clinical characteristics with COVID-19 test results. JAMA Network Open. 2020;**3**(9):e2019722. DOI: 10.1001/jamanetworkopen.2020.19722

[22] Ilie PC, Stefanescu S, Smith L. The role of vitamin D in the prevention of coronavirus disease 2019 infection and mortality. Aging Clinical and Experimental Research. 2020;**32**(7):1195-1198. DOI: 10.1007/s40520-020-01570-8

[23] D'Avolio A, Avataneo V, Manca A, et al. 25-Hydroxyvitamin D concentrations are lower in patients with positive PCR for SARS-CoV-2. Nutrients. 2020;**12**(5):1359. DOI: 10.3390/nu12051359

[24] Alcala-Diaz JF, Limia-Perez L, Gomez-Huelgas R, et al. Calcifediol treatment and hospital mortality due to COVID-19: A cohort study. Nutrients. 2021;**13**(6):1760. DOI: 10.3390/nu13061760

[25] Rastogi A, Bhansali A, Khare N, et al. Short term, high-dose vitamin D supplementation for COVID-19 disease: A randomised, placebo-controlled, study (SHADE study). Postgraduate Medical Journal. 2022;**98**(1156):87-90. DOI: 10.1136/postgradmedj-2020-139065

[26] Ding R, Xiao Z, Jiang Y, et al. Calcitriol ameliorates damage in high-salt diet-induced hypertension: Evidence of communication with the gut-kidney axis. Experimental Biology and Medicine (Maywood). 2022;**247**(8):624-640. DOI: 10.1177/15353702211062507

[27] Farid N, Rola N, Koch EAT, et al. Active vitamin D supplementation

and COVID-19 infections: Review. Irish Journal of Medical Science. 2021;**190**(4):1271-1274. DOI: 10.1007/s11845-020-02452-8

[28] Tylicki L, Puchalska-Reglińska E, Tylicki P, et al. Predictors of mortality in hemodialyzed patients after SARS-CoV-2 infection. Journal of Clinical Medicine. 2022;**11**(2):285. DOI: 10.3390/jcm11020285

[29] Zhang Y, Kong J, Deb DK, et al. Vitamin D receptor attenuates renal fibrosis by suppressing the renin-angiotensin system. Journal of The American Society of Nephrology. 2010;**21**(6):966-973. DOI: 10.1681/ASN.2009080872

[30] Li YC. Vitamin D regulation of the renin-angiotensin system. Journal of Cellular Biochemistry. 2003;**88**(2):327-331. DOI: 10.1002/jcb.10343

[31] Yuste C, Quiroga B, de Vinuesa SG, et al. The effect of some medications given to CKD patients on vitamin D levels. Nefrología. 2015;**35**(2):150-156. DOI: 10.1016/j.nefro.2015.05.016

[32] de Almeida LF, Coimbra TM. When less or more isn't enough: Renal maldevelopment arising from disequilibrium in the renin-angiotensin system. Frontiers in Pediatrics. 2019;**7**:296. DOI: 10.3389/fped.2019.00296

[33] Zhang Y, Deb DK, Kong J, et al. Long-term therapeutic effect of vitamin D analog doxercalciferol on diabetic nephropathy: Strong synergism with AT1 receptor antagonist. The American Journal of Physiology-Renal Physiology. 2009;**297**(3):F791-F801. DOI: 10.1152/ajprenal.00247.2009

[34] Zhang Z, Zhang Y, Ning G, et al. Combination therapy with AT1 blocker and vitamin D analog markedly ameliorates diabetic nephropathy: Blockade of compensatory renin increase. Proceedings of the National Academy of Sciences of the United States of America. 2008;**105**(41):15896-15901. DOI: 10.1073/pnas.0803751105

[35] Deng X, Cheng J, Shen M. Vitamin D improves diabetic nephropathy in rats by inhibiting renin and relieving oxidative stress. Journal of Endocrinological Investigation. 2016;**39**(6):657-666. DOI: 10.1007/s40618-015-0414-4

[36] Wang Y, Zhou J, Minto AW, et al. Altered vitamin D metabolism in type II diabetic mouse glomeruli may provide protection from diabetic nephropathy. Kidney International. 2006;**70**(5):882-891. DOI: 10.1038/sj.ki.5001624

[37] Zhang Y, Li XL, Sha NN, et al. Differential response of bone and kidney to ACEI in db/db mice: A potential effect of captopril on accelerating bone loss. Bone. 2017;**97**:222-232. DOI: 10.1016/j.bone.2017.01.029

[38] Li YC, Kong J, Wei M, et al. 1,25-Dihydroxyvitamin D(3) is a negative endocrine regulator of the renin-angiotensin system. The Journal of Clinical Investigation. 2002;**110**(2):229-238. DOI: 10.1172/JCI15219

[39] Li YC, Qiao G, Uskokovic M, et al. Vitamin D: A negative endocrine regulator of the renin-angiotensin system and blood pressure. The Journal of Steroid Biochemistry and Molecular Biology. 2004;**89-90**(1-5):387-392. DOI: 10.1016/j.jsbmb.2004.03.004

[40] Chandel N, Ayasolla K, Wen H, et al. Vitamin D receptor deficit induces activation of renin angiotensin system via SIRT1 modulation in podocytes. Experimental and Molecular Pathology. 2017;**102**(1):97-105. DOI: 10.1016/j.yexmp.2017.01.001

[41] Salhan D, Husain M, Subrati A, et al. HIV-induced kidney cell injury: Role of ROS-induced downregulated vitamin D receptor. The American Journal of Physiology-Renal Physiology. 2012;**303**(4):F503-F514. DOI: 10.1152/ajprenal.00170.2012

[42] Zhang Z, Sun L, Wang Y, et al. Renoprotective role of the vitamin D receptor in diabetic nephropathy. Kidney International. 2008;**73**(2):163-171. DOI: 10.1038/sj.ki.5002572

[43] Martínez-Arias L, Panizo S, Alonso-Montes C, et al. Effects of calcitriol and paricalcitol on renal fibrosis in CKD. Nephrology Dialysis Transplantation. 2021;**36**(5):793-803. DOI: 10.1093/ndt/gfaa373

[44] Tan X, Li Y, Liu Y. Paricalcitol attenuates renal interstitial fibrosis in obstructive nephropathy. Journal of The American Society of Nephrology. 2006;**17**(12):3382-3393. DOI: 10.1681/ASN.2006050520

[45] Kim CS, Joo SY, Lee KE, et al. Paricalcitol attenuates 4-hydroxy-2-hexenal-induced inflammation and epithelial-mesenchymal transition in human renal proximal tubular epithelial cells. PLoS One. 2013;**8**(5):e63186. DOI: 10.1371/journal.pone.0063186

[46] Jain SK, Parsanathan R, Levine SN, et al. The potential link between inherited G6PD deficiency, oxidative stress, and vitamin D deficiency and the racial inequities in mortality associated with COVID-19. Free Radical Biology and Medicine. 2020;**161**:84-91. DOI: 10.1016/j.freeradbiomed.2020.10.002

[47] Wu W, Liu D, Zhao Y, et al. Cholecalciferol pretreatment ameliorates ischemia/reperfusion-induced acute kidney injury through inhibiting ROS production, NF-κB pathway and pyroptosis. Acta Histochemica. 2022;**124**(4):151875. DOI: 10.1016/j.acthis.2022.151875

[48] Lee H, Lee H, Lim Y. Vitamin D3 improves lipophagy-associated renal lipid metabolism and tissue damage in diabetic mice. Nutrition Research. 2020;**80**:55-65. DOI: 10.1016/j.nutres.2020.06.007

[49] Khodir SA, Samaka RM, Ameen O. Autophagy and mTOR pathways mediate the potential renoprotective effects of vitamin D on diabetic nephropathy. International Journal of Nephrology. 2020;**2020**:7941861. DOI: 10.1155/2020/7941861

[50] Mitrašinović-Brulić M, Dervišević A, Začiragić A, et al. Vitamin D3 attenuates oxidative stress and regulates glucose level and leukocyte count in a semi-chronic streptozotocin-induced diabetes model. Journal of Diabetes & Metabolic Disorders. 2021;**20**(1):771-779. DOI: 10.1007/s40200-021-00814-2

[51] Campi I, Gennari L, Merlotti D, et al. Vitamin D and COVID-19 severity and related mortality: A prospective study in Italy. BMC Infectious Disease. 2021;**21**(1):566. DOI: 10.1186/s12879-021-06281-7

[52] Panichi V, Migliori M, Taccola D, et al. Effects of 1,25(OH)2D3 in experimental mesangial proliferative nephritis in rats. Kidney International. 2001;**60**(1):87-95. DOI: 10.1046/j.1523-1755.2001.00775.x

[53] Tan X, Wen X, Liu Y. Paricalcitol inhibits renal inflammation by promoting vitamin D receptor-mediated sequestration of NF-κB signaling. Journal of The American Society of Nephrology. 2008;**19**(9):1741-1752. DOI: 10.1681/ASN.2007060666

[54] Trivedi MK, Mondal S, Gangwar M, et al. Effects of cannabidiol interactions

with CYP2R1, CYP27B1, CYP24A1, and vitamin D3 receptors on spatial memory, pain, inflammation, and aging in vitamin D3 deficiency diet-induced rats. Cannabis and Cannabinoid Research. 19 Apr 2022. DOI: 10.1089/can.2021.0240

[55] Palmer MT, Lee YK, Maynard CL, et al. Lineage-specific effects of 1,25-dihydroxyvitamin D(3) on the development of effector CD4 T cells. Journal of Biological Chemistry. 2011;**286**(2):997-1004. DOI: 10.1074/jbc.M110.163790

[56] Talmor Y, Bernheim J, Klein O, et al. Calcitriol blunts pro-atherosclerotic parameters through NF-κB and p38 *in vitro*. European Journal of Clinical Investigation. 2008;**38**(8):548-554. DOI: 10.1111/j.1365-2362.2008.01977.x

[57] Tsoukas CD, Provvedini DM, Manolagas SC. 1,25-Dihydroxyvitamin D3: A novel immunoregulatory hormone. Science. 1984;**224**(4656):1438-1440. DOI: 10.1126/science.6427926

[58] LaFavers K. Disruption of kidney-immune system crosstalk in sepsis with acute kidney injury: Lessons learned from animal models and their application to human health. International Journal of Molecular Sciences. 2022;**23**(3):1702. DOI: 10.3390/ijms23031702

[59] Arora J, Wang J, Weaver V, et al. Novel insight into the role of the vitamin D receptor in the development and function of the immune system. The Journal of Steroid Biochemistry and Molecular Biology. 2022;**219**:106084. DOI: 10.1016/j.jsbmb.2022.106084

[60] Hsu CH, Patel SR, Young EW, et al. The biological action of calcitriol in renal failure. Kidney International. 1994;**46**(3):605-612. DOI: 10.1038/ki.1994.312

[61] Fraga M, Yáñez M, Sherman M, et al. Immunomodulation of T helper cells by tumor microenvironment in oral cancer is associated with CCR8 expression and rapid membrane vitamin D signaling pathway. Frontiers in Immunology. 2021;**12**:643298. DOI: 10.3389/fimmu.2021.643298

[62] Wang Y, Zheng J, Islam MS, et al. The role of $CD4^+$ $FoxP3^+$ regulatory T cells in the immunopathogenesis of COVID-19: Implications for treatment. International Journal of Biological Sciences. 2021;**17**(6):1507-1520. DOI: 10.7150/ijbs.59534

[63] Lamikanra AA, Tsang HP, Elsiddig S, et al. The migratory properties and numbers of T regulatory cell subsets in circulation are differentially influenced by season and are associated with vitamin D status. Frontiers in Immunology. 2020;**11**:685. DOI: 10.3389/fimmu.2020.00685

[64] Yuan D, Fang Z, Sun F, et al. Effect of vitamin D and tacrolimus combination therapy on IgA nephropathy. Medical Science Monitor. 2017;**23**:3170-3177. DOI: 10.12659/msm.905073

[65] Liu P, Li F, Xu X, et al. 1,25(OH)2D3 provides protection against diabetic kidney disease by downregulating the TLR4-MyD88-NF-κB pathway. Experimental and Molecular Pathology. 2020;**114**:104434. DOI: 10.1016/j.yexmp.2020.104434

[66] Penna G, Amuchastegui S, Cossetti C, et al. Treatment of experimental autoimmune prostatitis in nonobese diabetic mice by the vitamin D receptor agonist elocalcitol. The Journal of Immunology. 2006;**177**(12):8504-8511. DOI: 10.4049/jimmunol.177.12.8504

[67] Bock G, Prietl B, Mader JK, et al. The effect of vitamin D supplementation

on peripheral regulatory T cells and β cell function in healthy humans: A randomized controlled trial. Diabetes/ Metabolism Research and Reviews. 2011;**27**(8):942-945. DOI: 10.1002/ dmrr.1276

[68] Wang H, He X, Liang S, et al. Role of vitamin D in ulcerative colitis: An update on basic research and therapeutic applications. Expert Review of Gastroenterology and Hepatology. 2022;**16**(3):251-264. DOI: 10.1080/17474124.2022.2048817

[69] Ivanov II, McKenzie BS, Zhou L, et al. The orphan nuclear receptor RORgammat directs the differentiation program of proinflammatory IL-17+ T helper cells. Cell. 2006;**126**(6):1121-1133. DOI: 10.1016/j.cell.2006.07.035

[70] Heneka MT, Golenbock D, Latz E, et al. Immediate and long-term consequences of COVID-19 infections for the development of neurological disease. Alzheimer's Research & Therapy. 2020;**12**(1):69. DOI: 10.1186/ s13195-020-00640-3

[71] Fotuhi M, Mian A, Meysami S, et al. Neurobiology of COVID-19. Journal of Alzheimer's Disease. 2020;**76**(1):3-19. DOI: 10.3233/JAD-200581

[72] Mao L, Jin H, Wang M, et al. Neurologic manifestations of hospitalized patients with coronavirus disease 2019 in Wuhan, China. JAMA Neurology. 2020;**77**(6):683-690. DOI: 10.1001/jamaneurol.2020.1127

[73] Fasano A, Cereda E, Barichella M, et al. COVID-19 in Parkinson's disease patients living in lombardy, Italy. Movement Disorders. 2020;**35**(7):1089-1093. DOI: 10.1002/mds.28176

[74] Artusi CA, Romagnolo A, Ledda C, et al. COVID-19 and Parkinson's disease: What do we know so far? Journal of Parkinson's Disease. 2021;**11**(2):445-454. DOI: 10.3233/JPD-202463

[75] Morkavuk G, Demirkol A, Berber GE, et al. Comparison of dementia patients admission rates and dementia characteristics before and during the COVID-19 pandemic. Cureus. 2021;**13**(11):e19934. DOI: 10.7759/ cureus.19934

[76] Dilek TD, Boybay Z, Kologlu N, et al. The impact of SARS-CoV2 on the anxiety levels of subjects and on the anxiety and depression levels of their parents. Multiple Sclerosis and Related Disorders. 2021;**47**:102595. DOI: 10.1016/j. msard.2020.102595

[77] Aghagoli G, Gallo Marin B, Katchur NJ, et al. Neurological involvement in COVID-19 and potential mechanisms: A review. Neurocritical Care. 2021;**34**(3):1062-1071. DOI: 10.1007/s12028-020-01049-4

[78] Olivé-Cirera G, Fonseca E, Cantarín-Extremera V, et al. Impact of COVID-19 in immunosuppressed children with neuroimmunologic disorders. Neurology Neuroimmunology & Neuroinflammation. 2021;**9**(1):e1101. DOI: 10.1212/NXI.0000000000001101

[79] Chambergo-Michilot D, Barros-Sevillano S, Rivera-Torrejón O, et al. Factors associated with COVID-19 in people with Parkinson's disease: A systematic review and meta-analysis. European Journal of Neurology. 2021;**28**(10):3467-3477. DOI: 10.1111/ ene.14912

[80] Di Nicola M, Dattoli L, Moccia L, et al. Serum 25-hydroxyvitamin D levels and psychological distress symptoms in patients with affective disorders during the COVID-19 pandemic. Psychoneuroendocrinology.

2020;**122**:104869. DOI: 10.1016/j.psyneuen.2020.104869

[81] Serrano-Castro PJ, Estivill-Torrús G, Cabezudo-García P, et al. Impact of SARS-CoV-2 infection on neurodegenerative and neuropsychiatric diseases: A delayed pandemic? Neurología (English Edition). 2020;**35**(4):245-251. DOI: 10.1016/j.nrl.2020.04.002

[82] Hoffman MC, Freedman R, Law AJ, et al. Maternal nutrients and effects of gestational COVID-19 infection on fetal brain development. Clinical Nutrition ESPEN. 2021;**43**:1-8. DOI: 10.1016/j.clnesp.2021.04.019

[83] Xu Y, Baylink DJ, Chen CS, et al. The importance of vitamin D metabolism as a potential prophylactic, immunoregulatory and neuroprotective treatment for COVID-19. Journal of Translational Medicine. 2020;**18**(1):322. DOI: 10.1186/s12967-020-02488-5

[84] Menéndez SG, Martín Giménez VM, Holick MF, et al. COVID-19 and neurological sequelae: Vitamin D as a possible neuroprotective and/or neuroreparative agent. Life Sciences. 2022;**297**:120464. DOI: 10.1016/j.lfs.2022.120464

[85] Gomez-Pinedo U, Cuevas JA, Benito-Martín MS, et al. Vitamin D increases remyelination by promoting oligodendrocyte lineage differentiation. Brain and Behavior. 2020;**10**(1):e01498. DOI: 10.1002/brb3.1498

[86] Sabir MS, Haussler MR, Mallick S, et al. Optimal vitamin D spurs serotonin: 1,25-Dihydroxyvitamin D represses serotonin reuptake transport (SERT) and degradation (MAO-A) gene expression in cultured rat serotonergic neuronal cell lines. Genes and Nutrition. 2018;**13**:19. DOI: 10.1186/s12263-018-0605-7

[87] Cruz-López EO, Uijl E, Danser AHJ. Fifty years of research on the brain renin-angiotensin system: What have we learned? Clinical Science (Lond). 2021;**135**(14):1727-1731. DOI: 10.1042/CS20210579

[88] Malek MA. A brief review of interplay between vitamin D and angiotensin-converting enzyme 2: Implications for a potential treatment for COVID-19. Reviews in Medical Virology. 2020;**30**(5):e2119. DOI: 10.1002/rmv.2119

[89] Barrantes FJ. Central nervous system targets and routes for SARS-CoV-2: Current views and new hypotheses. ACS Chemical Neuroscience. 2020;**11**(18):2793-2803. DOI: 10.1021/acschemneuro.0c00434

[90] Labandeira-García JL, Garrido-Gil P, Rodriguez-Pallares J, et al. Brain renin-angiotensin system and dopaminergic cell vulnerability. Frontiers in Neuroanatomy. 2014;**8**:67. DOI: 10.3389/fnana.2014.00067

[91] Valenzuela R, Costa-Besada MA, Iglesias-Gonzalez J, et al. Mitochondrial angiotensin receptors in dopaminergic neurons. Role in cell protection and aging-related vulnerability to neurodegeneration. Cell Death & Disease. 2016;**7**(10):e2427. DOI: 10.1038/cddis.2016.327

[92] Uijl E, Ren L, Danser AHJ. Angiotensin generation in the brain: A re-evaluation. Clinical Science (Lond). 2018;**132**(8):839-850. DOI: 10.1042/CS20180236

[93] Passos-Silva DG, Verano-Braga T, Santos RA. Angiotensin-(1-7): Beyond the cardio-renal actions. Clinical Science (Lond). 2013;**124**(7):443-456. DOI: 10.1042/CS20120461

[94] Santos RA, Ferreira AJ, Verano-Braga T, et al. Angiotensin-converting

enzyme 2, angiotensin-(1-7) and mas: New players of the renin-angiotensin system. Journal of Endocrinology. 2013;**216**(2):R1-R17. DOI: 10.1530/JOE-12-0341

[95] de Barros Viana M, Rosário BDA, de Fátima Santana de Nazaré M, et al. COVID-19 in age-related neurodegenerative diseases: Is there a role for vitamin D3 as a possible therapeutic strategy? Annual Review of Neuroscience. 2020;**32**(2):235-247. DOI: 10.1515/revneuro-2020-0074

[96] Sha NN, Zhang JL, Poon CC, et al. Differential responses of bone to angiotensin II and angiotensin(1-7): Beneficial effects of ANG(1-7) on bone with exposure to high glucose. American Journal of Physiology-Endocrinology and Metabolism. 2021;**320**(1):E55-E70. DOI: 10.1152/ajpendo.00158.2020

[97] Wong TP, Ho KY, Ng EK, et al. Upregulation of ACE2-ANG-(1-7)-mas axis in jejunal enterocytes of type 1 diabetic rats: Implications for glucose transport. American Journal of Physiology-Endocrinology and Metabolism. 2012;**303**(5):E669-E681. DOI: 10.1152/ajpendo.00562.2011

[98] Villar-Cheda B, Dominguez-Meijide A, Valenzuela R, et al. Aging-related dysregulation of dopamine and angiotensin receptor interaction. Neurobiology of Aging. 2014;**35**(7):1726-1738. DOI: 10.1016/j.neurobiolaging.2014.01.017

[99] Diaz-Ruiz C, Villar-Cheda B, Dominguez-Meijide A, et al. Aging-related overactivity of the angiotensin/AT1 axis decreases Sirtuin 3 levels in the substantia nigra, which induces vulnerability to oxidative stress and neurodegeneration. The Journals of Gerontology. Series A-Biological Sciences and Medical Sciences. 2020;**75**(3):416-424. DOI: 10.1093/gerona/gly259

[100] Costa-Besada MA, Valenzuela R, Garrido-Gil P, et al. Paracrine and intracrine angiotensin 1-7/mas receptor axis in the substantia nigra of rodents, monkeys, and humans. Molecular Neurobiology. 2018;**55**(7):5847-5867. DOI: 10.1007/s12035-017-0805-y

[101] Getachew B, Tizabi Y. Vitamin D and COVID-19: Role of ACE2, age, gender, and ethnicity. Journal of Medical Virology. 2021;**93**(9):5285-5294. DOI: 10.1002/jmv.27075

[102] Peng MY, Liu WC, Zheng JQ, et al. Immunological aspects of SARS-CoV-2 infection and the putative beneficial role of vitamin-D. International Journal of Molecular Sciences. 2021;**22**(10):5251. DOI: 10.3390/ijms22105251

[103] Mansur JL, Tajer C, Mariani J, et al. Vitamin D high doses supplementation could represent a promising alternative to prevent or treat COVID-19 infection. Clínica e Investigación en Arteriosclerosis. 2020;**32**(6):267-277. DOI: 10.1016/j.arteri.2020.05.003

[104] He MC, Shi Z, Sha NN, et al. Paricalcitol alleviates lipopolysaccharide-induced depressive-like behavior by suppressing hypothalamic microglia activation and neuroinflammation. Biochemical Pharmacology. 2019;**163**:1-8. DOI: 10.1016/j.bcp.2019.01.021

[105] Feng R, He MC, Li Q, et al. Phenol glycosides extract of Fructus Ligustri Lucidi attenuated depressive-like behaviors by suppressing neuroinflammation in hypothalamus of mice. Phytotherapy Research. 2020;**34**(12):3273-3286. DOI: 10.1002/ptr.6777